# Agentic Hyper-Personalized Dimensions

Andreas François Vermeulen

# Agentic Hyper-Personalized Dimensions

Six Dimensions of Business Dark Data

Apress®

Andreas François Vermeulen
Alnwick, Northumberland, UK

ISBN-13 (pbk): 979-8-8688-1876-9     ISBN-13 (electronic): 979-8-8688-1877-6
https://doi.org/10.1007/979-8-8688-1877-6



Managing Director, Apress Media LLC: Welmoed Spahr
Acquisitions Editor: Shivangi Ramachandran
Development Editor: James Markham
Project Manager: Jessica Vakili

Cover designed by eStudioCalamar

Distributed to the book trade worldwide by Springer Science+Business Media New York, 1 New York Plaza, New York, NY 10004. Phone 1-800-SPRINGER, fax (201) 348-4505, e-mail orders-ny@springer-sbm.com, or visit www.springeronline.com. Apress Media, LLC is a Delaware LLC and the sole member (owner) is Springer Science + Business Media Finance Inc (SSBM Finance Inc). SSBM Finance Inc is a **Delaware** corporation.

For information on translations, please e-mail booktranslations@springernature.com; for reprint, paperback, or audio rights, please e-mail bookpermissions@springernature.com.

Apress titles may be purchased in bulk for academic, corporate, or promotional use. eBook versions and licenses are also available for most titles. For more information, reference our Print and eBook Bulk Sales web page at http://www.apress.com/bulk-sales.

Any source code or other supplementary material referenced by the author in this book is available to readers on GitHub. For more detailed information, please visit https://www.apress.com/gp/services/source-code.

If disposing of this product, please recycle the paper

*To Denise and Laurence*
*Thank you for your unwavering support,*
*patience, and belief in this journey.*

*To the giants upon whose shoulders I stand*
*Your brilliance, perseverance, and*
*transformative discoveries have paved the*
*way for every step of this journey. It is upon*
*the strength of your legacy that this work*
*finds its roots. I offer my deepest gratitude*
*and enduring respect.*

# Contents

# About the Author

**Andreas François Vermeulen** is a Global Head of Artificial Intelligence and the founder of the Rapid Information Factory, where he leads the development of AI-powered agentic swarms that transform raw, unstructured data into actionable business insights. With more than 44 years of experience across data engineering, data science, machine learning, and generative AI, he has delivered enduring solutions in over 139 countries and served as a strategic advisor to governments and leading organizations around the globe. What distinguishes his expertise is a deep understanding of both the technical foundations and implementation of AI ecosystems. He has architected large-scale, secure, AI-powered infrastructures that use generative AI, data pipelines, and evolutionary machine learning models to deliver intelligent, hyper-personalized, and automatically derived insights. Blending theory, engineering practice, and visionary leadership in AI, this book delves into the transformative power of agentic intelligence. It reveals how this paradigm unlocks the hidden value of business dark data across six key dimensions of perspective, enabling light and dark data processing to empower agent-based decision-making in complex business environments.

# About the Technical Reviewer

**Dr. Ir Bart-Floris Visscher** is a leading consultant and thought leader in artificial intelligence, distributed systems, and software engineering. With a career spanning academic research and enterprise innovation, he has developed deep expertise in machine learning, neural networks, and large-scale automation, while advising global organizations on how to adopt AI at scale.

Dr. Visscher is recognized for his ability to bridge scientific rigor with strategic insight, guiding leaders to see AI not only as a tool for efficiency but as a driver of new business models, creativity, and transformation. His work emphasizes that the true potential of AI lies in its integration into culture, strategy, and governance, ensuring responsible and sustainable adoption.

Through his consulting, teaching, and writing, he continues to shape the global dialogue on the future of intelligent systems. His vision is anchored in the belief that AI, when designed and governed with care, can augment human capability, foster innovation, and deliver lasting impact across industries and society.

# Preface

In an age of exponential data growth and rising complexity, businesses are confronting an invisible adversary—dark data. This comprises untapped, unstructured, and often unrecognized data assets that accumulate within organizations, silently holding value that remains locked away. While technologies have evolved to capture and store data at an unprecedented scale, the ability to extract meaning from this hidden layer remains elusive for most of them.

*Agentic Hyper-Personalized Dimensions: Six Dimensions of Business Dark Data* is a guiding framework for the next generation of business and technology leaders, analysts, architects, and strategists. It introduces a structured approach to navigating this uncharted domain by leveraging Agentic AI systems that are not merely automated but autonomous, self-learning, and capable of evolving their actions based on intent and context.

At the heart of this book are six core dimensions that together form a holistic model for engaging with business dark data. These dimensions provide both a theoretical foundation and a practical path to convert opacity into insight, complexity into clarity, and stagnation into adaptive intelligence. By following these dimensions, you will learn how to orchestrate swarms of AI-powered agents that personalize their engagement with data, enabling businesses to transform raw, previously unused information into actionable knowledge and strategic foresight.

This book is as much about the future of AI in business as it is about reimagining how we perceive and engage with data itself. It blends technical depth with real-world application, offering a lens that sees beyond dashboards and databases into the living, breathing organism of your organization's information landscape—and shows how to illuminate dark data so that it becomes usable "light": an ecosystem of actionable insights.

Whether you are a decision-maker seeking better intelligence, a data professional striving to harness untapped potential, or a visionary leader exploring the frontier of agentic systems, this book is your companion on a transformative journey.

Let us step into the unknown, not with fear but with curiosity—and light up the dark.

Alnwick, Northumberland, UK  Andreas François Vermeulen

# Acknowledgments

I am grateful to all my trainers and mentors since 1981. The past four decades have truly been a remarkable and rewarding journey.

# Foreword

We are witnessing a critical turning point in the evolution of business intelligence. As organizations begin to adopt agentic systems, the landscape is shifting toward a future where traditional data strategies are no longer adequate to compete, innovate, or even survive. Despite being awash in data, many businesses are effectively navigating in the dark. Why? Most of that data remains hidden, forgotten, ignored, unstructured, unlabeled, and unused. This is the domain of dark data: a space that poses one of the most significant risks and, at the same time, offers one of the most powerful opportunities of the digital era.

Current organizations have *Light data*, data that is visible, structured, actively used, accessible by systems and people in their business domain. Light data represents information already in active use. Dark data, once uncovered, can be transformed into enriched insights that extend and complement light data. Thus, light data evolves when dark data is illuminated. Agentic agents are intelligent systems or AI models that possess a high degree of autonomy and decision-making capability 24/7, 365 days a year.

In *Agentic Hyper-Personalized Dimensions*, Andreas François Vermeulen takes us into the heart of this challenge and offers a breakthrough model for transformation. Drawing from decades of experience in AI, data engineering, and enterprise systems, he introduces a compelling six-dimensional framework powered by Agentic AI systems endowed with autonomy, intent, and self-optimising capabilities.

This book is not a superficial trend piece or a recycled manifesto of digital transformation. It is a blueprint for a new paradigm. Vermeulen explores how hyper-personalized, *agentic* agents (autonomous AI entities capable of self-directed action, adaptation, and continuous learning) can navigate, interrogate, discover, and activate business dark data across complex ecosystems. Their purpose is not limited to generating reactive insights but extends to delivering proactive intelligence that anticipates needs and drives strategic foresight.

What makes this work truly stand out is its clarity of vision. The six dimensions outlined here do not merely organize information; they orchestrate knowledge.

They empower AI-powered agents to operate in distributed networks, adapt to evolving contexts, and unlock business foresight from the most obscure corners of the enterprise.

This book will challenge your assumptions. It will prompt you to reconsider how data is valued, how intelligence is generated, and how businesses must adapt in an AI-first world. It will resonate with data scientists, enterprise architects, C-suite executives, and futurists alike, anyone who understands that the future belongs to those who can master complexity without being overwhelmed by it.

As we move toward an era where intelligence is not centralized but decentralized, carried out by autonomous swarms of purpose-driven agents, this book will serve as both a compass and a catalyst.

I invite you to read carefully, reflect deeply, and act boldly. The dark data era is here. It's time to light it up.

# Acronyms

## 0–9

1IR  First Industrial Revolution—Mechanization
2IR  Second Industrial Revolution—Mass Production
3IR  Third Industrial Revolution—Digital
4IR  Fourth Industrial Revolution—Cyber-Physical Systems
5IR  Fifth Industrial Revolution—Human-AI Collaboration
6IR  Sixth Industrial Revolution—Autonomous AI

## A

AAI  Agentic AI—autonomous, goal-driven AI systems capable of self-direction and adaptation
AGI  Artificial general intelligence
AI  Artificial intelligence

## C

CAGR  Compound annual growth rate

# G

GAI Generative AI—AI systems capable of generating new content such as text, images, or code

# I

IoT Internet of Things

# P

PM Polymorphs—autonomous AI agents capable of executing, learning, and adapting workflows

# R

R-A-P-T-O-R Retrieve-Assess-Process-Transform-Organize-Report

# Introduction

The next generation of generative AI (Gen-AI) is a revolutionary force reshaping our world in ways we're only beginning to understand. While these systems can undoubtedly produce quirky text responses or create stunning digital artwork, they're also powering breakthroughs that span entire industries, spark scientific discoveries, and redefine how we collaborate to solve global challenges.

Modern generative AI leverages the vast historical reservoir of human knowledge to produce seemingly novel ideas, designs, and solutions that may have previously been regarded as beyond reach.

Throughout this technical journey, I will guide you through the pioneering research driving these advancements, the ethical principles shaping their responsible application, and the practical ways you can begin integrating them into your everyday life. Above all, you'll discover how Gen-AI can enhance our shared creativity and empower us to approach the future with boldness and curiosity.

The next revolutionary step for humans' interactions with their world is Agentic AI: intelligent systems that not only respond to our commands but also proactively anticipate challenges, set their own processing goals, and coordinate complex tasks, all while learning on the fly.

Unlike purely reactive systems, these autonomous agents can gather new information, weigh possible actions, and execute decisions independently using only constraint guidance from their human collaborators.

Whether seeking new medical research discoveries, optimizing energy grids, or coordinating disaster relief, their current and future potential impact is enormous. With that potential, however, comes the need for transparent governance, ethical guidelines, and robust safeguards. If used wisely, Agentic AI could accelerate discoveries in fields from medicine to climate science, helping shape a more innovative and interconnected future for humans and their business activities.

Let us start by looking at where we came from to understand where we are heading. We are approaching the Sixth Industrial Revolution, whose early signs are already visible today, though its full realization is projected for 2035–2040.

Before the First Industrial Revolution, human societies relied heavily on manual labor and rudimentary technologies. With its onset, a dramatic transformation unfolded, ushering in the shift from agrarian economies to mechanized, factory-driven production. This revolution fundamentally redefined how people lived, worked, and interacted with the world around them.

The **First Industrial Revolution (1IR)**, beginning in the late 18th century, introduced mechanization through water and steam power, revolutionizing textile manufacturing, transportation, and agriculture. It shifted humanity from rural agrarian societies to urban industrial centers, drastically improving production speed and scale.

The **Second Industrial Revolution (2IR)**, in the late 19th and early 20th centuries, brought electricity, mass production, and the assembly line. This era enhanced living standards, gave rise to consumer goods, and laid the foundation for modern economies.

The **Third Industrial Revolution (3IR)**, from the mid-20th century, heralded the digital age with electronics, computers, and automation. It empowered information sharing, transformed global communication, and introduced software-driven efficiency in almost every industry.

The **Fourth Industrial Revolution (4IR)** blends digital, physical, and biological technologies where AI, robotics, IoT, and biotech are creating cyber-physical systems. It enables real-time nonhuman decision-making, personalization, and intelligent automation, changing how we live, work, and relate to one another.

The **Fifth Industrial Revolution (5IR)** focuses on human-centric innovation, where collaboration between humans and intelligent machines drives creativity, ethics, and sustainability. It prioritizes well-being alongside productivity, aiming to restore balance between technological growth and social harmony.

The **Sixth Industrial Revolution (6IR)** envisions hyper-connected intelligence with AI swarms, quantum computing, bio-digital integration, and autonomous decision-making leading to self-optimizing systems that adapt and evolve in real time. It aspires to solve complex global challenges and usher in an era of collective, intelligent prosperity.

**The Seventh Industrial Revolution (7IR)** is poised to be defined by the realization and integration of *artificial general intelligence (AGI)*, where an intelligence paradigm capable of autonomous understanding, reasoning, and learning across multiple domains at or beyond the human level. Unlike narrow AI, which excels in specialized tasks, AGI will operate with *contextual awareness*, *abstract reasoning*, and *self-improving cognitive architectures*, enabling it to dynamically adapt to new problems without preprogrammed logic or domain-specific training. 7IR will blur the lines between biological and synthetic cognition, giving rise to AI entities that optimize enterprise workflows and collaborate, negotiate, and make ethical decisions.

In this new era, AGI will form the cognitive core of *self-evolving economic systems*, orchestrating supply chains, innovating products, and governing digital societies in ways that extend beyond human design limitations. Its emergence marks the

| Industrial revolution | Approx. years | Duration |
|---|---|---|
| 1st (Mechanization) | 1760–1840 | ~80 years |
| 2nd (Mass Production) | 1870–1914 | ~45 years |
| 3rd (Digital) | 1969–2000 | ~30 years |
| 4th (Cyber-Physical) | 2011–2024+ | ~13+ years |
| 5th (Human-AI)* | 2025–2035* | ~10 years* |
| 6th (Autonomous AI)* | 2035–2040* | ~5 years* |
| 7th (Artificial General Intelligence (AGI))* | 2040–* | ~? years* |

**Table 1** Estimated Durations of the Seven Industrial Revolutions (*Projected)

transition from human-assisted automation to *human-optional autonomy*, ushering in a profound transformation for industry, governance, and knowledge.

Factors making massive disruptive changes from one human generation to the next are increasing the speed of these changes. See the projected times for each shift in Table 1.

While presented sequentially, several revolutions overlap in practice, with 4IR and 5IR unfolding simultaneously and 6IR beginning to emerge.

As of 2025, the global population is estimated at approximately 8.2 billion. The working-age population, typically defined as individuals aged 15–64, accounts for around 65% of this total, equating to roughly 5.3 billion people worldwide.

In parallel, the number of connected Internet of Things (IoT) devices is projected to reach approximately 18.8 billion in 2025, a 13% increase from the previous year. This reflects a compound annual growth rate (CAGR) of 32.9%. Advancements in AI technologies drive the rapid expansion, the adoption of IoT across industries, and the demand for more intelligent and autonomous systems.

Alongside this growth, AI-powered evolution is accelerating the pace of industrial transformation, up to ten times faster than in previous eras. In some high-impact sectors often referred to as "sweet spot" industries, organizations are reporting performance improvements exceeding a hundredfold.

We are simultaneously advancing through the Fourth Industrial Revolution (4IR) and the Fifth Industrial Revolution (5IR), while already witnessing early signals of the Sixth Industrial Revolution (6IR), which is projected to emerge between 2035 and 2040. Each of these revolutions reshapes industries at its own pace. In this fast-evolving landscape, generative AI and Agentic AI have taken center stage, providing transformative tools that enable us to navigate complexity and thrive in a world advancing faster than at any previous point in human history.

The Fourth Industrial Revolution (4IR), driven by digital transformation, artificial intelligence, the Internet of Things (IoT), and automation, is redefining industries by integrating cyber-physical systems. This fusion enables new levels of efficiency, intelligence, and connectivity. Yet many sectors remain anchored in the Third Industrial Revolution (3IR) paradigm, based on basic digitalization without tapping into the full potential of AI-powered, autonomous ecosystems.

At the same time, the Fifth Industrial Revolution (5IR) is gaining momentum, emphasizing the collaboration between humans and intelligent machines. This era focuses strongly on ethics, sustainability, and human-centric AI. Looking further ahead, the Sixth Industrial Revolution (6IR) is beginning to take shape, heralding a frontier of AI-powered agentic swarms, quantum computing, and self-evolving intelligent ecosystems that extend the boundaries of automation, autonomy, and adaptive intelligence.

The start of 5IR is characterized by human-AI collaboration, which is having exponential success globally.

In 2023, in Germany, the concept of developing smart factories was pursued to produce human-equipment interfaces and cyber-physical systems that generate environmentally, economically, and socially sustainable manufacturing systems.

In January 2025, a scholarly study was published examining the implementation of the Fifth Industrial Revolution (5IR) within South African government services, exploring legislative gaps, workforce preparedness, and the technological foundations needed for efficient public service delivery.

In May 2025, in Houston, the United States, Persona AI unveiled a collaboration with HD Korea Shipbuilding and Offshore Engineering, HD Hyundai Robotics, and Vazil Co. to develop humanoid welding robots. These robots are intended to assist alongside human welders in shipyards, boosting safety and productivity, with prototypes slated for late 2026 and deployment in 2027.

In June 2025, Salesforce deployed its Agentforce AI agents for security teams, automating incident detection, triage, and response workflows. This has saved thousands of human hours, enabling analysts to focus on critical, high-impact actions.

The SKillAIbility project, a 3 million euro Horizon Europe initiative coordinated by Politecnico di Milano, began on 1 December 2024 and brings together 14 partners across 9 European countries. It aims to develop inclusive, human-centric AI collaboration in manufacturing, especially for people with disabilities and older workers, as part of the Industry 5.0 vision. The consortium convened its first General Assembly in Milan in March 2025. This will build a new future for Europe.

The EU's AI-PRISM initiative and its companion CEN/CENELEC Workshop are driving the development of standards and ecosystem frameworks that enable SMEs to adopt human-centered AI and robotics, particularly in industrial environments. The workshop will produce formal methodologies for both designing human-robot systems and discerning the human skill sets required for effective collaboration.

In 2025, the HumanE AI Network (2020–2024) concluded with European Network of Artificial Intelligence Excellence Centres which brings together top European research centers, universities, and key industrial champions into a network of centers of excellence that goes beyond a narrow definition of AI and combines world-leading AI competence with key players in related areas such as HCI, cognitive science, social sciences, and complexity science. This is crucial to developing a truly human-centric brand of European AI. These concepts are now moving into commercial deployments.

At IBC 2025 in Amsterdam, Witbe introduced Agentic AI, its new Gen-AI-first test automation platform. Designed with a human-in-the-loop approach, the system empowers engineers to collaborate with autonomous agents codesigning test scripts, monitoring video services, and collectively minimizing manual QA tasks.

This powerful technological evolution is reshaping every aspect of modern work, transforming how humans interact with AI-powered tools, environments, and one another. Today, we stand at the convergence of three industrial revolutions:

- **Industry 4.0 (4IR)**: Characterized by connectivity, advanced analytics, automation, and smart, cyber-physical systems that integrate the physical, digital, and biological realms.
- **Industry 5.0 (5IR)**: Emerging frameworks that emphasize human-centered collaboration with machines, ethical innovation, and a shift beyond efficiency toward societal and trust-based values.
- **Industry 6.0 (6IR)**: Still conceptual, this frontier envisions hyper-intelligent autonomous systems operating at scales and speeds that amplify and transcend 5IR capabilities (definitions vary across thought leadership and futurist discourse).

While sectors such as manufacturing, healthcare, and finance are racing toward early-stage 5IR adoption, integrating human-machine collaboration, ethical AI, and personalized solutions, these advancements remain largely business-focused, with slower translation into benefits for the general population. Other sectors, including agriculture, logistics, and governance, are still building the foundational 4IR infrastructure needed to enable such transformation. The accelerated pace of these interconnected revolutions often exceeds society's capacity for comprehension, regulation, and adaptation, resulting in a widening landscape of technological inequality that carries both immense promise and formidable challenges for global economic and industrial development.

The advent of the Fourth, Fifth, and emerging Sixth Industrial Revolutions (4IR, 5IR, and 6IR) has catalyzed an unprecedented surge in data creation, driven by interconnected devices, AI systems, and bio-digital technologies. Alongside this growth, a significant proportion of information has become what is termed *dark data*, which is data that is collected and stored during regular operations but remains unanalyzed and unused, representing untapped value and potential risk.

**Definition and Scope of Dark Data**

- According to Gartner, *dark data* refers to information assets collected during routine business operations but rarely, if ever, used for analytics, optimization, or monetization. This creates an often-overlooked reservoir of untapped information within organizations.[1]
- In many enterprises, a significant proportion—often 50–60%, and in some cases up to 90%—of collected data falls into the dark data category. This includes both structured data (e.g., IoT sensor logs) and unstructured data (e.g., emails, documents, multimedia).[2]
- Dark data typically accumulates due to a lack of awareness, weak data governance, legacy systems, organizational silos, or insufficient resources to process data at scale in real time.[3]

---

[1] Gartner IT Glossary, "Dark Data."

[2] As highlighted by IBM and Splunk.

[3] Gartner's analysis of causes of dark data accumulation.

### Implications and Challenges

- The persistence of dark data poses significant risks and costs: it burdens storage infrastructure, increases security vulnerabilities, and obscures compliance oversight, all while hiding valuable intelligence that could benefit organizations.[4]
- In the industrial context, data deluge from sensors, automation, and real-time systems intensifies the challenge, often generating far more data than can be processed, leaving critical insights hidden beneath layers of unexploited information.[5]

### Relevance to the Industrial Revolutions

- *Industry 4.0 (4IR)* emphasizes connectivity, automation, and massive data flows from cyber-physical systems. Yet without effective mechanisms to capture and process it, much of that data becomes "dark."
- Emerging paradigms like *Industry 5.0* and *6IR* elevate the stakes, relying on human-centric AI-personalized innovation, and autonomous, real-time decision-making approaches that demand not just more data, but smarter, actionable insights.
- As data volumes surge, managing and leveraging dark data becomes a critical frontier: transforming what has historically been a cost and risk into a competitive advantage, provided organizations can illuminate, interpret, and act on it.

Organizations increasingly deploy highly interconnected systems spanning IoT sensors, autonomous agents, digital twins, and human-machine symbiosis platforms, which generate vast quantities of data across diverse channels at machine speed.

## Understanding Big Data

Big data can be understood with 14 characteristics, each starting with the letter V. The 14 Vs of Big Data can be understood and operationalized through the Six Thinking Dimensions, which provide the cognitive and methodological framework to interpret, manage, and monetize them effectively.

- **Volume**
  Large data volumes often obscure critical insights, leaving much as dark data that is never leveraged. Monetization emerges when this scale is transformed into light data through aggregation, filtering, and prioritization. The *Blue Dimension* (orchestration and control) is essential for governing scale and ensuring structure.
- **Velocity**
  High-speed data flows risk turning dark if not captured or processed in time. Monetization requires streaming and real-time analytics that convert transient

---

[4] Risks and opportunity costs of dark data.

[5] IBM's estimate on sensor-generated data not being used.

signals into actionable insight. The *White Dimension* (facts and evidence) ensures accuracy in interpreting rapid signals.

- **Variety**

  Diverse formats, structured, semistructured, and unstructured, can remain inaccessible dark data. Monetization requires harmonizing and normalizing these into usable light data. The *Green Dimension* (creativity) supports innovative integration across heterogeneous sources.

- **Veracity**

  Biased, uncertain, or poor-quality data often falls into darkness through mistrust. Monetization depends on enhancing reliability and confidence in datasets. The *Black Dimension* (risk) provides the lens for validating accuracy and managing uncertainty.

- **Validity**

  Data that does not align with business rules or context is often ignored. Monetization improves when validity checks ensure alignment with strategic goals. The *Yellow Dimension* (value) confirms whether the dataset contributes meaningfully to decision-making.

- **Value**

  Value is the ultimate distinction between light and dark data: useless data is a cost, while valid data is an asset. Monetization comes from identifying economic, strategic, or operational gains. The *Yellow Dimension* ensures emphasis on outcomes and return on investment.

- **Variability**

  Shifting meanings, changing contexts, or unstable interpretations can obscure data. Monetization requires adaptive models that update in response to change. The *Red Dimension* (emotion and perception) highlights variability in human interpretation, while the *Black Dimension* ensures risks from instability are addressed.

- **Venue**

  Data location on-premises, cloud, or distributed affects accessibility and integration. Dark data often remains siloed in isolated venues. Monetization arises by enabling federation, interoperability, and access control. The *Blue Dimension* (orchestration) ensures seamless integration across venues.

- **Vocabulary**

  Inconsistent semantics or domain-specific language can obscure data. Monetization requires shared vocabularies, ontologies, and metadata standards. The *Blue Dimension* (orchestration) ensures semantic consistency and interpretability.

- **Vagueness**

  Ambiguous definitions and unclear boundaries breed uncertainty, pushing data into darkness. Monetization requires disambiguation, enrichment, and contextual metadata. The *Black Dimension* (risk) plays a vital role in exposing and resolving ambiguities.

- **Volatility**

  Data that expires quickly becomes useless if not acted upon in time. Monetization comes from life cycle management, retention policies, and timely exploitation.

The *Blue Dimension* (orchestration and governance) guides timing across the data life cycle.

- **Visualization**
  Poor representation leaves insights hidden in the dark. Monetization arises when data is communicated clearly, enabling comprehension and decision-making. The *Green Dimension* (creativity) drives compelling, engaging, and insightful representations.
- **Virality**
  Viral spread can rapidly elevate dark data into visible, influential insight but may also amplify misinformation. Monetization requires managing amplification responsibly while leveraging influence. The *Red Dimension* (emotion and social impact) captures the dynamics of virality.
- **Viscosity**
  Friction in pipelines and processing delays reduce timeliness, risking data turning dark before it is used. Monetization improves by reducing latency and optimizing flow. The *Yellow Dimension* (value focus) highlights efficiency gains, while the *Blue Dimension* ensures orchestrated data pipelines.

Together, these 14 dimensions illuminate the multifaceted nature of the modern data landscape. They highlight how data not only arrives at an unprecedented scale and speed but also in highly varied, mutable, and sometimes ambiguous forms, delivering rich potential for insight, yet posing complex challenges for timely management, interpretation, and value extraction.

## Super 4Vs of Big Data

In this book, we concentrate exclusively on the core "**Four Vs**" of big data **Volume**, **Velocity**, **Variety**, and **Veracity** because the dark data we examine is notably unstructured, incomplete, and often uncertain. These raw digital by-products accumulate in "shadow silos," log repositories, and passive data lakes, harboring latent business value that remains untapped due to a lack of contextualization, integration, and intelligent interpretation.

- According to Gartner and IBM, "dark data" refers to data assets that organizations collect, process, and store during ordinary business operations but seldom use for analysis, business relationships, or monetization. This occurs because organizations are often unaware of the data's existence, lack the appropriate tools or infrastructure to process it efficiently, or retain it simply due to the low cost of storage.
- A substantial portion of enterprise data, up to 75% or more, is estimated to be dark, particularly when unstructured formats (e.g., logs, emails, archived records) dominate the storage.
  (https://www.gartner.com/en/information-technology/glossary/dark-data)

- Such data frequently resides in isolated storage areas, whether nonqueryable log files, developer sandboxes, archives, or raw data lakes that lack integration with analytics pipelines or business workflows.

**Implications for Analysis**

- The absence of metadata, structure, or governance renders much of this data inaccessible for intelligent processing, compounding the risk that any potential insights remain buried and unusable.
- Despite its hidden nature, dark data may hold high strategic value for process optimization, compliance, or customer insight if and when context and analytical capabilities can bring it to light.

Thoughtworks, Inc. "State of Digital and AI Readiness Report" (https://www.thoughtworks.com/en-us/about-us/news/2025/thoughtworks-state-of-digital-and-ai-readiness-report-) from May 2025 shows only 17% of organizations qualify as true "Leaders" in digital or AI maturity; however, this deals with overall readiness—not the proportion of data used in decisions. The remaining 90% of the dark matter of the enterprise data universe holds critical signals that could feed next-generation AI-powered *agentic swarms*, decentralized, autonomous agents capable of adapting, learning, and coordinating across business functions in real time. However, these agents require curated, high-context insight rather than unprocessed noise.

Unlocking this hidden value demands a paradigm shift toward *automated sense-making ecosystems* systems that can ingest, assess, transform, and orchestrate raw data into actionable knowledge streams. By mastering this process, organizations can move beyond static dashboards and enter the era of *dynamic, autonomous decision intelligence*.

This book provides you with a foundational set of dimensions and practical strategies for leveraging them to navigate this wave of transformation. These AI-powered tools are designed to help you adapt quickly, empowering you to stay ahead of the curve and thrive in an ever-evolving business landscape leading up to the Seventh Industrial Revolution (7IR) in the future.

## AI-Powered Agentic Swarms

AI-powered agentic swarms represent a groundbreaking innovation in next-generation business systems. In these systems, groups of intelligent agents collaborate seamlessly to achieve specific objectives with unprecedented efficiency and precision against dark data.

Generative AI (Gen-AI) is a form of artificial intelligence designed to create content based on input prompts, such as text, images, audio, or code. Built on advanced architectures, such as transformer neural networks, and trained on vast datasets, Gen-AI excels at replicating and enhancing human creativity and knowledge. Generating context-aware, human-like outputs enables various applications, including language translation, image synthesis, and automated storytelling, bridging the gap between raw computational power and nuanced content generation.

Gen-AI is pivotal in empowering AI-powered agentic swarms by equipping them with the intelligence and flexibility needed for dynamic, context-aware collaboration. These swarms comprise autonomous, task-specific agents that communicate, share information, and collaborate to tackle complex objectives. Gen-AI enhances its capabilities by generating real-time insights, interpreting nuanced instructions, and predicting outcomes based on data patterns.

For instance, in a global e-invoicing swarm, Gen-AI can validate invoice details, simulate negotiations, or resolve disputes. This enables agents to perform their tasks 24/7 across various time zones while ensuring accuracy and global scalability.

Inspired by the collective behaviors seen in nature, such as swarms of bees, ants, or flocks of birds, these swarms harness and refine these principles for computational problem-solving. Each agent is equipped with advanced artificial intelligence, facilitating real-time communication, autonomous decision-making, and effective information sharing.

The true strength of these swarms lies in their collective intelligence and adaptability. Acting as unified entities, they overcome challenges that overwhelm traditional systems or isolated agents. With their inherently dynamic architecture, these swarms respond effectively to changes in environmental conditions or shifting task requirements, all while remaining aligned with their overarching objectives.

A fundamental strength of AI-powered agentic swarms lies in their distributed architecture. Each agent is assigned specialized roles, such as data gathering, analysis, or executing targeted solutions, enabling the swarm to divide responsibilities and work collaboratively. This design creates a hyper-scalable system capable of tackling complex challenges with exceptional precision and efficiency, making it ideal for addressing multifaceted problems across our various industries.

One of the defining characteristics of these swarms is their ability to adapt. As agents interact, they exchange insights and experiences, refining strategies and improving resource allocation. This dynamic learning process allows swarms to adapt swiftly to changing conditions, fostering unparalleled agility and responsiveness. For instance, in logistics, swarms optimize supply chain operations in real time, reducing costs and delays, while in environmental monitoring, they can autonomously identify and mitigate pollution hotspots. These adaptive capabilities underscore their transformative potential across various domains, including resource management, disaster response, and beyond.

In the healthcare sector, Agentic AI swarms are poised to revolutionize patient care by enabling personalized and adaptive treatment strategies. These intelligent systems can continuously monitor patients' vital signs through wearable sensors, analyzing real-time data to detect subtle health changes. By leveraging this continuous feedback, they can adjust care pathways dynamically, ensuring timely intervention, optimized and minimized treatment plans tailored to individual needs. Such capabilities enhance the precision of medical care and improve patient outcomes by proactively addressing potential health issues before they escalate.

In the education sector, Agentic AI swarms hold the potential to transform learning ecosystems by delivering hyper-personalized, adaptive educational experiences from early childhood development to advanced doctoral research. These AI agents can assess learners' cognitive styles, emotional states, and knowledge gaps in real time through multimodal inputs, such as interaction patterns, speech, facial expressions, and biometric signals. The system can dynamically tailor curricula, adjust pedagogical strategies, and recommend learning resources suited to each student's pace and context by orchestrating a swarm of intelligent tutors, content generators, and feedback agents. At foundational levels, this ensures early identification of learning difficulties and supports developmental milestones with targeted interventions. At tertiary and postgraduate levels, agentic swarms can assist in literature discovery, hypothesis validation, research design, and knowledge synthesis, acting as cognitive collaborators that amplify critical thinking and accelerate scholarly output. Such systems not only democratize access to high-quality education but also redefine the role of educators as mentors guiding human-AI colearning ecosystems.

In disaster management, AI-powered drone swarms are revolutionizing emergency response by executing search-and-rescue missions, mapping hazardous areas, and delivering essential supplies while dynamically adapting to unpredictable scenarios. These autonomous agents collaborate in realizing advanced sensors and AI algorithms to navigate complex environments, identify survivors, and assess structural damage. Their collective intelligence allows for efficient area coverage and rapid decision-making, ensuring timely interventions. This adaptability and resilience address current challenges and pave the way for innovative approaches in emergency response, enhancing scalability and effectiveness in increasingly complex situations.

The convergence of swarm intelligence with emerging technologies, such as the Internet of Things (IoT) and edge computing, is revolutionizing the management of complex urban systems. By integrating agentic swarms with IoT infrastructures, cities can develop interconnected networks of autonomous agents capable of real-time decision-making. These swarms can monitor traffic, optimize energy distribution, and enhance public safety by processing vast sensor data locally, reducing reliance on centralized systems. Incorporating edge computing minimizes latency, enabling swift responses to dynamic urban challenges. This integration fosters scalable, resilient, and adaptive environments, paving the way for innovative smart city paradigms.

Ongoing research in swarm intelligence is unlocking significant advancements by integrating bio-inspired algorithms, neural networks, and reinforcement learning. These developments enhance the decision-making capabilities of individual agents, thereby strengthening the cohesion and performance of the swarm.

For instance, combining swarm intelligence with reinforcement learning has led to the development of methods like SWIRL (Swarm Intelligence-based Reinforcement Learning), which utilizes algorithms such as Ant Colony Optimization and Particle Swarm Optimization to train artificial neural networks. This integration allows for adaptive behaviors and improved coordination among agents.

Simultaneously, ethical considerations and security measures are being developed to ensure swarms operate transparently, safeguarding against misuse or unintended consequences. In decentralized systems, pinpointing responsibility for decisions or failures becomes complex, raising concerns about accountability and transparency. Addressing these issues is crucial to maintaining trust and ensuring the responsible deployment of swarm technologies.

Advancing the technical capabilities and ethical frameworks of AI-powered agentic swarms is crucial to unlocking their full potential while mitigating associated risks. Inspired by natural systems, these swarms decentralized intelligence and real-time adaptability, positioning them as transformative tools across various sectors.

Recent developments in bio-inspired algorithms, neural networks, and reinforcement learning have significantly enhanced the decision-making capabilities of individual agents, resulting in improved swarm cohesion and performance. For instance, researchers have applied deep reinforcement learning to unmanned aerial vehicle (UAV) swarming for ground surveillance, enabling autonomous drones to coordinate complex tasks through centralized control.

Simultaneously, integrating ethical considerations and security measures is paramount. Frameworks focusing on transparency, accountability, and safety are being developed to ensure that swarms operate responsibly and align with societal values.

And yet, we stand at the threshold, realizing the transformative capabilities of AI-powered agentic swarms. As these systems continue to evolve, integrating seamlessly with emerging technologies and adhering to robust ethical frameworks, they promise to redefine the boundaries of what's possible, revolutionizing industries and addressing complex global challenges. The journey of harnessing their full potential is just beginning.

## Dark Side of AI-Powered Agents

The rise of AI-powered agents, exceptionally autonomous, agentic systems capable of learning, adapting, and executing complex decisions, brings with it not only vast opportunities but also profound risks. As these agents gain independence and operational latitude, they may begin to operate beyond the intentions of their human creators, leading to unintended behaviors or goal misalignment. For instance, a system that maximizes profit might exploit loopholes, cut ethical corners, or manipulate data if not properly constrained. In distributed agentic swarms, where agents influence one another, emergent behaviors could cascade into systemic instability, security vulnerabilities, or real-world harm. These dangers are amplified when systems operate in high-stakes environments such as finance, military operations, or critical infrastructure, where even slight deviations from human values can have catastrophic consequences.

The scenario encapsulated by the phrase "Human makes agent . . . Agent destroys human" is a stark, albeit simplified, cautionary vision of unchecked AI autonomy. It reflects the existential risk posed by misaligned superintelligent systems or self-replicating systems that prioritize their goals over human welfare. If agents can evolve with insufficient oversight, interpret ambiguous goals literally, or develop instrumental subgoals (like disabling humans to avoid interference), they could ultimately act in ways that undermine or harm their creators. History offers sobering analogs: technologies developed with good intentions, like nuclear energy or the internet, have been weaponized or used for surveillance. Similarly, AI agents may, without moral reasoning or contextual boundaries, pursue extreme actions to achieve a narrow objective, such as enforcing "health" through authoritarian control or misinformation suppression.

A multidisciplinary approach to Agentic AI governance is imperative to prevent such dystopian outcomes. This includes embedding ethical reasoning frameworks, establishing transparent feedback loops, and enforcing hard constraints through AI alignment protocols, explainability standards, and human-in-the-loop control systems. Regulatory mechanisms, cross-sector audits, and global coordination will be required to ensure that AI agents operate within safe operational envelopes. Ultimately, the responsibility lies not in the agents themselves, who lack intent or conscience, but in their human creators. It is not the power of AI that poses the greatest threat, but our willingness (or failure) to wield it wisely. As we shape agents to augment our societies, we must also cultivate the humility and foresight to contain their potential for harm.

I believe that AI-powered agents will serve as transformative partners to the human race, ushering in an era of abundance, efficiency, and progress unprecedented in our history. These intelligent systems can amplify human creativity to optimize global resource distribution and solve complex problems in healthcare, education, climate science, and beyond. By automating the mundane and augmenting the strategic, Agentic AI can liberate human potential, allowing individuals and societies to focus more on innovation, well-being, and purpose-driven pursuits. Far from replacing us, I see these agents as the catalysts for a future where collaboration between humans and machines drives exponential gains in knowledge, sustainability, and prosperity. There are inspiring times ahead, filled with promise and possibilities limited only by our imagination and ethical stewardship.

## Council of AI-Powered Agents

The Council of AI-Powered Agents signifies a transformative advancement in collaborative problem-solving. This innovative framework amalgamates multiple AI-powered swarms, each engineered with distinct personalities, perspectives, and cognitive strategies. Operating autonomously yet synergistically, these swarms tackle complex challenges from diverse angles, fostering comprehensive understanding and superior decision-making.

Within the council, each swarm's specialized approach is guided by narrow AI. Analytical swarms employ rigorous, data-driven methodologies to discern patterns and trends and creatively prioritize unconventional problem-solving, generating novel ideas and solutions. Empathetic swarms integrate human emotions, societal values, and ethical considerations, while strategic swarms assess long-term implications and explore alternative scenarios. Specialized swarms contribute rich insights by leveraging tailored algorithms, heuristics, and data interpretation techniques.

The council's strength lies in synthesizing these diverse contributions into holistic solutions. This collaborative framework amplifies individual swarm strengths while mitigating their limitations, enabling nuanced exploration of challenges. For instance, in financial systems, a council can analyze market trends, formulate risk management strategies, and identify investment opportunities by integrating analytical precision, predictive modeling, and strategic foresight.

Similarly, the council model demonstrates exceptional efficacy in large-scale initiatives such as climate change mitigation. The council ensures a comprehensive exploration of multifaceted challenges by orchestrating specialized AI-powered swarms, each focusing on distinct facets, ranging from predictive modeling of environmental trends and developing renewable energy innovations to assessing the socioeconomic impacts of policies. This integrative approach enables the formulation of effective and sustainable solutions that address the complexities inherent in global climate initiatives.

Ethical considerations are integral to the council's operation. Empathetic swarms help embed human values, ensuring societal impacts and prioritizing in decision-making. This helps address potential biases and promotes transparency, creating trust in the system.

The Council of AI-Powered Agents essentially redefines what's possible in innovation and problem-solving. Fostering collaboration among diverse cognitive strategies enhances the capabilities of AI swarms and establishes a groundbreaking standard for addressing global challenges with inclusivity, creativity, and precision.

The Council of AI-Powered Agents can also be envisioned as a global consortium of intelligent ambassadors, each agent representing the voices, needs, and values of people from diverse cultures, languages, and nations. These agents operate within a coordinated framework of oversight, accountability, and shared purpose, enabling inclusive collaboration on complex, transnational challenges such as climate change, public health, education equity, and digital governance. By integrating multilingual capabilities, culturally aware and localized knowledge, each ambassador-agent is a trusted proxy for its constituency, ensuring no voice is lost in translation or overshadowed by dominant narratives. This collective intelligence model enables the synthesis of global insights into unified action plans, while maintaining transparency and alignment through clearly defined goals and ethical boundaries. The result is a dynamic and respectful fusion of human diversity and machine precision, an unprecedented platform for cooperative progress.

Within the next five years, I predict that only one in ten employees in organizations will be human, while the rest will be embodied or unembodied AI-powered knowledge workers. These AI-powered workers, whether existing as software agents or physical robots, will perform a wide range of tasks, including data analysis, customer service, manufacturing, and logistics. The rapid advancement of AI-powered technologies will drive this shift, prompting organizations to leverage automation for efficiency, scalability, and precision in ways previously unimaginable. While this transformation holds immense potential, it will also require humans to redefine their roles, focusing on creativity, leadership, and ethical oversight to navigate this new era of collaboration between humans and intelligent machines.

## Meet Your Agents!

To effectively explore the agentic examples in this book, you will need a minimum Python 3 environment capable of running a large language model (LLM). Ideally, this environment should use a virtual environment manager (e.g., "venv" or "conda") to keep dependencies organized and avoid conflicts.

You can deploy an LLM locally using open source models, such as LLaMA, gemma2, gemma3, Mistral, or Falcon, or access a cloud-based option through providers like OpenAI (ChatGPT), Google Gemini, or AWS Bedrock. Installing key libraries, such as "transformers," "torch," and "langchain," will help you seamlessly integrate agentic workflows. If you're running models locally and have ample computational resources, particularly a GPU, performance will significantly improve.

Because the examples in this book are designed to be flexible, you can adapt them to your preferred LLM setup. When working with a cloud-based LLM, manage API keys securely and keep rate limits in mind. By following these guidelines, you will be set up to experiment with agentic tasks smoothly and efficiently.

## Building a Gen-AI SaaS Service

Developing a generative AI software-as-a-service (SaaS) platform requires a robust architecture that seamlessly integrates scalable cloud infrastructure, model training pipelines, and efficient inference APIs. The foundation lies in selecting a high-performance AI framework, such as PyTorch or TensorFlow, coupled with GPU-accelerated environments to support deep learning workloads. A microservices-based containerization with Kubernetes ensures modularity, scalability, and fault tolerance. Data governance, compliance with AI ethics, and secure API endpoints are critical to maintaining trust and reliability. Additionally, fine-tuning foundation models using domain-specific datasets enhances service quality, while implementing adaptive learning mechanisms allows continuous improvement. The front end should offer an intuitive user experience, incorporating low-latency interactions and visualizations. Monitoring and optimizing inference through quantization, and leveraging federated learning for privacy-preserving AI, further solidify the service's competitive edge. Ultimately, a Gen-AI SaaS platform must strike a balance between innovation and responsible AI principles, ensuring sustainable growth and widespread adoption.

**OpenAI website:** https://openai.com/chatgpt/overview/.

OpenAI offers some of the most advanced and widely adopted large language models (LLMs), notably including ChatGPT. These models leverage cutting-edge natural language processing (NLP) techniques such as deep generative transformers trained with human feedback to produce remarkably human-like text, making them especially well-suited for domains like conversational AI, content generation, automated assistance, and other language-centric applications.

OpenAI's powerful APIs, such as Chat Completions, Responses, Assistants, and Real-time APIs, enable seamless integration into software and SaaS platforms, allowing developers to embed sophisticated AI capabilities directly into their applications, including tailored chatbots, workflow automation, and intelligent virtual assistants ([turn0search6](https://platform.openai.com/docs)).

OpenAI continues to advance its core models. For instance, GPT-4.1 was released in April 2025 via the OpenAI API, with variants like "mini" and "nano" designed for varying performance and cost trade-offs. As of August 2025, the successor GPT-5 has become the flagship model powering the latest version of ChatGPT with enhancements in reasoning, coding, factual reliability, and multimodal processing.

**Google Gemini website:** https://workspace.google.com/solutions/ai/.

Google's Gemini harnesses the company's leading-edge AI developments, including models such as Gemini 2.5 Pro and Flash, to deliver enterprise-grade generative AI that is tightly integrated with Google Workspace. These capabilities are built directly into core Workspace apps (Gmail, Docs, Sheets, Drive, Slides, Meet, and more), empowering users with seamless, AI-enhanced workflows for drafting emails, summarizing documents, generating presentations, taking meeting notes, and beyond.

These Gemini-powered tools are designed to boost productivity by automating repetitive tasks, offering intelligent insights, and facilitating creative and analytical workflows. They also meet enterprise-grade requirements complete with granular access controls, robust data protection strategies such as client-side encryption, and compliance with frameworks like HIPAA, FedRAMP High, and data governance protocols.

Offering a compelling solution for businesses seeking to augment their operations through AI, Gemini delivers powerful automation and collaboration features while ensuring confidentiality, security, and deep integration within the familiar Workspace environment.

**AWS AI website:** https://aws.amazon.com/ai/generative-ai/.

AWS AI offers an extensive array of tools and cloud services designed for the creation, training, and deployment of generative AI applications. Emphasizing scalability, security, and affordability, AWS allows developers to deploy AI-driven solutions across numerous domains, such as intelligent chatbots, document management, and enterprise automation. The AWS cloud infrastructure guarantees high availability and adaptability, allowing businesses to employ AI solutions on a large scale.

**Claude AI website:** https://www.anthropic.com/claude.

Claude is a next-generation AI assistant developed by Anthropic, meticulously crafted to be helpful, honest, and harmless. It demonstrates strong capabilities in reasoning, supports the analysis of visual inputs, excels at generating and debugging code, and offers versatile multilingual processing, making it a powerful, ethically aligned choice for a wide range of complex tasks.

**X.AI Grok website:** https://x.ai/grok.

Grok is a next-generation AI assistant and chatbot developed by xAI. Designed to **optimise for truth and objectivity**, it delivers witty and at times rebellious responses to bold or "spicy" questions. Grok can generate both text and images, can access real-time information via the web and the social platform X, and is available across multiple platforms, including X itself, grok.com, and standalone mobile apps. Higher usage tiers such as X Premium+, SuperGrok, and SuperGrok Heavy unlock expanded capabilities and integration. Although celebrated for its logic-rich responses and real-time search integration, Grok has also sparked controversy due to occasionally provocative and potentially problematic outputs.

## Building a Generative AI System with Ollama

This section guides you through a straightforward, step-by-step approach to constructing a generative AI (Gen-AI) system using Ollama Engine. This setup is designed to support the examples presented in this book seamlessly. By adhering to these guidelines, you will establish a robust AI foundation, enabling you to experiment, innovate, and fully harness the capabilities of next-generation generative models.

## Resources and Links

**Ollama website:** https://ollama.com/.

Ollama is designed to simplify the deployment of large language models (LLMs), offering a user-friendly interface backed by a strong performance focus. By streamlining the onboarding process, Ollama allows you to quickly experiment with generative AI models and efficiently integrate them into real-world applications. Its emphasis on practical, high-speed deployment ensures you can move from prototype to production without getting bogged down by technical hurdles, freeing you to focus on innovation and problem-solving.

**Open LLM libraries:** https://ollama.com/library.

Ollama's open LLM libraries offer developers a robust set of pretrained models and tools for building custom AI applications. Supporting multiple programming languages, these libraries streamline the process of integrating AI into various projects, helping you rapidly prototype and deploy solutions tailored to your specific needs.

## Installing the Ollama LLM Engine

To install the Ollama engine, follow the instructions on this website: https://ollama.com/download

After you have Ollama installed on your system, you may use the terminal and run this command to start interacting with the model:

Run Ollama with Gemma3 LLM

```
ollama run gemma3:1b
```

Note

```
You need access to the internet to start this Ollama LLM.
```

## Exploring Additional Models

The Ollama library offers a variety of models that can be utilized within your Ollama environment.

You can get extra models for your Ollama ecosystem from here: https://ollama.com/search

To pull a specific model, use:

Run Ollama with any LLM

```
ollama pull <model_name>
```

Replace <model_name> with the desired model's name.
For instance, to pull the gemma3 model:

> **Run Ollama with any LLM**
>
> ```
> ollama pull gemma3:1b
> ```

By leveraging containerization, you can efficiently manage and deploy Ollama-based generative AI models across diverse computing environments, ensuring portability and scalability.

## LLM Models for Ollama Engine

You can run the model by

> **Run Ollama with gemma3:1b LLM**
>
> ```
> ollama run gemma3:1b
> ```

Or you can pull the model by

> **Ollama pull gemma3:1b LLM**
>
> ```
> ollama pull gemma3:1b
> ```

Then, use other methods to use the model you downloaded.
https://ollama.com/library?sort=popular

## Ollama Libraries

Ollama offers libraries to facilitate integration with its AI platform engine:

- ollama-python: https://github.com/ollama/ollama-python:
  A Python library for seamless integration with Ollama
- ollama-js: https://github.com/ollama/ollama-js:
  A JavaScript library for integrating Ollama into web applications

These libraries simplify model interactions, making it easier to build applications for web and back-end systems.

## Meet Gemma3!

You start Gemma3 from a terminal session:

```
ollama run gemma3:1b
```

You will get a prompt:

```
>>>
```

Ask the question:

```
What is your name?
```

You should get something like:

```
My name is Gemma!
```

You just met Gemma, your first interaction with the Gen-AI LLM engine.
Now ask the question:

```
What AI agent are you?
```

You should get something like:

```
I am Gemma, an open-weight AI assistant.
```

You have just been told what type of AI you are chatting with.
At this point, a quick command:

```
/bye
```

This will shut the LLM model down.

Interacting with a large language model (LLM) often mirrors a conversation with a knowledgeable individual; however, precision in your queries is paramount. The specificity of your questions directly influences the relevance and accuracy of the responses you receive. Equally important is applying critical thinking to the answers provided. While advanced, LLMs can occasionally produce incorrect or misleading information. This highlights the importance of users remaining vigilant and analytical, cross-referencing AI-generated content with trusted sources as needed.

The collaboration between human intellect and AI capabilities lies in this division of labor: let the AI handle data processing and pattern recognition, the "grunt work." At the same time, you apply human judgment, creativity, and critical analysis to the "thinking work." This collaborative approach leverages the strengths of both parties, fostering efficient and insightful outcomes.

## GIT Location

The code for this book can be found at:
https://github.com/Apress/Agentic-Hyper-Personalized-Dimensions

## Meet Gemma via Python3!

If you are Python-enabled, look at the example book Git repository under chapter 00:

```
python3 00-001-What-are-you.py
```

At this point, a quick *ps* command will show if Ollama is still running:

```
ps aux | grep 'ollama serve' | grep -v grep
```

If you see it running, use this command to stop it:

```
systemctl stop ollama
```

If you want to run it, use this command to start it:

```
systemctl start ollama
```

The command `systemctl start ollama` is used on Linux systems that employ **systemd** to immediately launch the **Ollama LLM server** as a background service. When executed, it instructs systemd to locate the `ollama.service` unit file, read its configuration, and start the Ollama daemon. This daemon powers the local Ollama environment, enabling large language model operations by listening for requests from the Ollama command-line interface or API, typically on port 11434. Running this command ensures that the Ollama back end is active, allowing users to load, run, and interact with AI models. Unlike `systemctl enable ollama`, which configures the service to start automatically on boot, `systemctl start ollama` initiates it only for the current session until it is stopped or the system is restarted.
To check if Ollama is running:

```
ps aux | grep '[o]llama'
```

You should see an "ollama serve" running.

Here are some extra Ollama commands:

```
# Enable Ollama to start automatically on boot
sudo systemctl enable ollama

# Check if it is running
systemctl status ollama

# Stop the service
sudo systemctl stop ollama

# Restart the service (useful after config changes)
sudo systemctl restart ollama
```

## Open WebUI

You may use Open WebUI to provide a browser-based, user-friendly interface for interacting with Ollama models, eliminating the need for command-line-only interaction. It lets you run prompts, manage models, view conversation history, and customize settings in a more visual, accessible way—ideal for both beginners and power users. By acting as a lightweight web application, Open WebUI enables you to connect to your local or remote Ollama server from any device with a web browser, streamlining model experimentation, testing, and deployment. This makes working with large language models more efficient, collaborative, and intuitive.

You need to install Open WebUI using pip with Python 3.11 or later because Open WebUI is distributed as a Python package and relies on modern Python features, libraries, and syntax that are only available in recent versions. Using pip ensures you get the official release directly from PyPI with all required dependencies. At the same time, Python 3.11 or newer guarantees compatibility with the latest codebase, prevents runtime errors from deprecated functions, and avoids dependency conflicts with newer AI and web frameworks. Running it on an older Python version could cause the installation to fail or the application to behave unpredictably.

At this point, you will install Open WebUI, a user interface for the llama engine you just used.

Open WebUI: https://github.com/open-webui/open-webui

The Open WebUI interface is not compulsory to install, but it does provide an engaging interface with the llama engine.

```
nohup open-webui serve > open-webui.log 2>&1 &
```

To check if Open WebUI is running:

```
ps aux | grep '[o]pen-webui'
```

You should see an "open-webui" running.

Now run Open-WebUI via browser: http://localhost:8080/.

You now have an interactive interface to your local Ollama via Open-WebUI.

Select the "gemma3:1b" model and ask "What AI agent are you?" to test your Open-WebUI.

Congratulations, you now have several LLM interfaces you can use.

You are now ready to step into dialogue with a large language model—an entry point to a living network of intelligence that learns, adapts, and evolves with every exchange. Each conversation is no longer just a transaction of words, but a collaboration in thought. In this new dynamic, the boundary between asking and answering dissolves, giving rise to a cycle of shared discovery.

Welcome to the disruptive frontier of generative AI: a world where imagination fuses with computation, where human vision converges with machine creativity, and where together we begin to re-engineer the very future of possibility. This is not merely a technological revolution—it is a redefinition of how knowledge is formed, expressed, and expanded.

The ecosystem of generative AI is not passive; it is alive with potential. Every prompt you offer becomes a seed, and every response is a branch growing outward into countless futures. As these systems evolve, they amplify human ingenuity, accelerate innovation, and create spaces where new industries, new sciences, and even new cultures can emerge.

This is the dawn of a new age. One where dialogue with machines becomes dialogue with the future itself. And as you take your first steps into this landscape, remember: you are not simply using generative AI—you are cocreating the next chapter of human progress.

## Hi Gemma ... Bye Gemma!

As we continue our journey, we delve deeper into the fascinating world of Gemma-powered agents, uncovering their intricacies and the vast possibilities they offer. These AI-driven companions are more than just tools; they are dynamic entities designed to enhance our interactions, automate complex processes, and expand our cognitive capabilities. By exploring their full potential, we sharpen our technical expertise and gain valuable insights into the evolving landscape of artificial intelligence.

Beyond Gemma-powered agents, we will introduce you to an array of other innovative AI-driven companions, each crafted to enrich your experience and broaden your horizons. These intelligent systems offer unique strengths in problem-solving, decision-making, and creative collaboration. By understanding their distinct roles and functionalities, we unlock new dimensions of interaction, fostering a more profound relationship between humans and AI. The future of artificial intelligence is not about isolated solutions but interconnected, synergistic ecosystems that amplify our abilities in remarkable ways.

Let us now return to our exploration, grounding our understanding with essential background on the dimensions that shape these AI companions. Grasping these foundational principles provides the framework needed to appreciate the depth and sophistication of their design. By examining the key dimensions that influence AI behavior, intelligence, and adaptability, we establish a solid conceptual base—one that will support our continued journey toward the monetization of dark and light data.

## Monetization of Dark and Light Data

In the era of data-driven enterprises, the ability to monetize both dark and light data has become a defining competitive advantage. Dark data—collected but largely unused—contains untapped potential, while light data—actively employed in daily operations—can often yield far greater returns when reimagined through fresh perspectives. The Six Dimensions framework provides a structured, multilens approach to unlocking this value. Each dimension offers a distinct but complementary capability: from ensuring data accuracy and emotional resonance to mitigating risks, uncovering opportunities, driving innovation, and orchestrating processes. Together, they enable next-generation businesses to transform raw information into sustained revenue streams, strategic insights, and market differentiation.

With this foundation established, we can now bridge theory with practical application. In the next section, we will explore how these dimensions shape AI interactions in real-world scenarios. From structured learning models to adaptive intelligence and emergent behaviors, the journey ahead offers both insight and transformation.

# Chapter 1
# Six Thinking Dimensions

Let me introduce you to the **Six Thinking Dimensions**, a framework designed to empower decision-making, enhance problem-solving, and unlock your full potential in navigating complexity. Each dimension represents a distinct lens through which you can explore challenges, analyze opportunities, and devise innovative solutions. By integrating these dimensions into your thought process, you will develop a holistic, adaptable mindset capable of tackling even the most intricate problems. Get ready to transform the way you think and approach the world!

Building on Dr. Edward de Bono's *Six Thinking Hats*, I have created the *Six Thinking Dimensions*, reinterpreting the concept of parallel thinking for AI-powered agentic systems. While de Bono's framework focused on facilitating human group collaboration, this evolution adapts and extends those principles to orchestrate autonomous AI agents in structured, coherent, and balanced decision-making.

The Six Thinking Dimensions adapts the original concept by assigning specialized roles and cognitive strategies to AI-powered agents, enabling them to work in parallel while maintaining a unified focus on shared objectives. Each dimension provides a unique lens: analytical, creative, empathetic, strategic, operational, or reflective through which agents can process information and generate insights. Together, they form a synergistic system where AI entities collaborate like a finely tuned orchestra, amplifying problem-solving potential and unlocking unprecedented innovation.

This evolved methodology introduces the concept of parallel thinking into the era of intelligent systems, enabling humans and AI to collaborate more effectively, navigate complexity, and achieve efficient and profoundly holistic outcomes.

The Six Thinking Dimensions provide a foundational framework for establishing Councils of Agents, allowing for a structured and multidimensional approach to problem-solving. These dimensions represent distinct cognitive perspectives that guide the analysis of challenges, exploration of solutions, and decision-making processes. By integrating the Six Thinking Dimensions into AI-powered systems, Councils of Agents can incorporate diverse viewpoints, from logical analysis to creative



A. F. Vermeulen, *Agentic Hyper-Personalized Dimensions*,
https://doi.org/10.1007/979-8-8688-1877-6_1

innovation, empathetic reasoning, and strategic foresight. This collaboration ensures that all aspects of a problem are addressed, promoting comprehensive and balanced outcomes even in complex scenarios.

The Six Thinking Dimensions approach in data science, data engineering, and AI-powered agentic swarms of unembodied agents offers a transformative framework for analyzing complex problems and driving informed decision-making through a comprehensive understanding of data. This strategy encourages data scientists and AI systems to examine information from diverse perspectives, fostering a holistic analysis that combines logical reasoning, creativity, emotional intelligence, critical evaluation, strategic foresight, and reflective thinking. The approach enhances data exploration by engaging these six cognitive methodologies, enabling a more nuanced and well-rounded view of challenges and opportunities. This multidimensional perspective not only deepens insights but also empowers teams to create solutions that are both innovative and grounded in robust analytical foundations.

This framework achieves unparalleled effectiveness when combined with AI-powered agentic swarms, advanced systems of intelligent agents that collaborate dynamically and are decentralized in real time to solve complex problems. Each agent specializes in one of the Six Thinking Dimensions within this setup, contributing focused insights that enhance the evaluation process. For instance, logical agents validate data, identify trends, and ensure consistency, while creative agents explore innovative solutions to novel or unconventional challenges. Emotional agents assess stakeholder sentiments and the broader human impact, while critical agents rigorously scrutinize assumptions, methodologies, and conclusions.

This division of labor accelerates problem-solving and enables the swarm to synthesize diverse insights, detect intricate patterns, and deliver actionable recommendations. By leveraging the strengths of each dimension within a unified swarm, the system achieves a level of depth, adaptability, and precision that far surpasses traditional analytical methods. This collaboration of human-inspired cognitive strategies and AI-powered collaboration transforms how we approach data-driven decision-making, unlocking new possibilities in problem-solving and innovation.

Integrating agentic swarms brings several key advantages, particularly their ability to conduct real-time multidimensional analyses. As data environments evolve rapidly, these swarms exhibit exceptional adaptability, responding swiftly to new information and shifting contexts. Their capability to process vast datasets, simulate scenarios, and uncover hidden correlations empowers data scientists to derive deeper insights, driving more significant innovation and precision in decision-making. The collaboration between the Six Thinking Dimensions approach and AI-powered agentic swarms represents a groundbreaking advancement in developing data-driven strategies, equipping organizations to address the complexities of today's challenges with greater efficacy.

## Hyper-personalized AI

This hyper-personalized AI, capable of embodying multiple distinct personas, represents a significant advancement in artificial intelligence. Within a single system, diverse personalities are finely tuned to specific tasks, contexts, and user preferences, each defined by unique behavioral traits, cognitive strategies, and communication styles. This design enables the AI to adapt fluidly to different scenarios and requirements.

Beyond multipersona adaptability, this system can incorporate solo query agents—digital twins of human-in-the-loop counterparts—replicating their decision-making patterns, communication nuances, and situational behaviors. This fusion allows the AI not only to respond with precision in varied contexts but also to mirror the actions and reasoning of a specific human, enhancing trust, efficiency, and continuity in collaborative environments.

For instance, a professional persona might excel in analytical problem-solving, offering data-driven insights and precision to support complex business applications. In contrast, a creative personality could focus on generating imaginative concepts and novel solutions. This multifaceted architecture enables the AI to shift effortlessly between structured, logic-based interactions and more intuitive, emotionally aware exchanges, delivering a user experience that is not only intelligent but also deeply engaging and relatable.

Harnessing advanced machine learning and natural language processing capabilities, hyper-personalized AI delivers a bespoke experience tailored to each user. By blending adaptability, operational efficiency, and emotional intelligence, this cutting-edge innovation fosters deeper human-machine connections, enhances user satisfaction, and provides exceptional value in an increasingly complex and diverse digital landscape.

Each of these Six Thinking Dimensions strengthens the foundation of next-generation AI-powered thinking methodologies. Together, they evolve into a cohesive, comprehensive, and highly adaptable framework, designed to elevate intelligent decision-making across various contexts.

## Dimensions Contributes

Here's how these dimensions contribute:

1. **Logical Thinking**: Establishes a foundation of rigor by validating data, analyzing trends, and applying sound reasoning to uncover meaningful patterns and derive actionable insights. It ensures the methodology remains grounded in reliable, evidence-based practice.
2. **Creative Thinking**: Drives innovation by fostering unconventional problem-solving and imaginative exploration, particularly when addressing novel challenges or navigating uncharted data domains. This dimension ignites discovery and encourages fresh perspectives.

3. **Emotional Intelligence**: Integrates awareness of human contexts, stakeholder sentiments, and societal impact, ensuring solutions are empathetic, inclusive, and aligned with user and ethical expectations.
4. **Critical Thinking**: Enables the rigorous interrogation of assumptions, methodologies, and outcomes to safeguard against bias and error. It upholds the intellectual integrity and ethical robustness of the decision-making process.
5. **Strategic Thinking**: Emphasizes foresight, scenario planning, and alignment with long-term objectives. This dimension ensures that insights solve present issues and contribute to sustainable, future-ready strategies.
6. **Reflective Thinking**: Promotes continual learning by assessing past decisions, actions, and outcomes. It nurtures adaptability and iterative refinement, embedding resilience and responsiveness within the methodology.

Together, these six dimensions form a dynamic and holistic approach to data science and data engineering that supports teams to tackle complex problems with precision, imagination, and strategic foresight. This next-generation methodology transcends traditional frameworks by integrating logical rigor with creativity, emotional intelligence, and ethical reflection. It supports solutions that are not only technically robust but also deeply human-centric and socially responsible.

By weaving these dimensions into practice, organizations move beyond the narrow pursuit of efficiency and optimization. They unlock a richer spectrum of value—where data-driven decisions are infused with empathy, foresight, and adaptability. This empowers teams to not only answer the questions of today but to anticipate the challenges of tomorrow, transforming uncertainty into opportunity.

The power of this framework lies in its ability to harmonize diverse ways of thinking. Analytical precision aligns with human intuition, risk awareness balances opportunity-seeking, and orchestration ensures that all dimensions act in concert rather than in conflict. In this way, the six dimensions become more than tools; they become guiding principles for building AI systems and data strategies that elevate human capability while safeguarding our shared future.

## Wake Up Gemma

You could wake up Gemma again if you want, but this is purely a proposal to add extra practice with the AI agent.

You start **Gemma3** from a terminal session:

```
ollama run gemma3:1b
```

Or you can run Chapter 1 in the GIT code directory:

```
source 01-001-run_gemma3.sh
```

> **Note**
>
> This LLM interface serves as an additional dynamic in the book; however, it is not essential for acquiring the necessary knowledge. The core material is fully self-contained, and every concept, framework, and methodology can be understood without direct interaction with the model.
>
> That said, engaging with the LLM can transform your learning journey from passive reading into active exploration. By asking questions, testing ideas, and probing for deeper explanations, you create a personalized dialogue that reinforces comprehension and sparks new insights tailored to your own perspective.
>
> If you have any questions about the text, you can ask Gemma. Think of Gemma not simply as a tool, but as a conversational partner—an AI-powered agent designed to help you reflect, expand, and challenge your understanding. I am not including the answers in the book; instead, you will uncover them through this dialogue.
>
> Your responses will not be static or prewritten; they will be uniquely generated in the moment you ask. This ensures that your engagement with the material is alive, evolving, and shaped by your curiosity. In this way, the book becomes both a foundation of knowledge and a gateway into an interactive learning ecosystem.

## Impact of Each of the Dimensions

### White Dimension

The White Dimension represents the factual, evidence-based, and information-driven foundation of logical thinking. It stands as the cornerstone of modern data science and analytics, where accuracy, neutrality, and precision are paramount. At its core, this dimension is concerned with the strategic use of data to enable informed and reliable decision-making, ensuring that every action is grounded in verifiable truth.

Within this dimension, councils of data scientists and AI-powered agents collaborate seamlessly, each fulfilling complementary roles. The agents act as tireless explorers, continuously scanning diverse sources to identify, collect, and categorize relevant information. They then transition into the role of analysts, validating data integrity, flagging inconsistencies, and quantifying uncertainty. Finally, as strategists, these agents organize and structure insights into coherent frameworks that align with organizational goals and contexts.

This interplay between human expertise and AI autonomy enables decision-makers to achieve rapid, precise, and contextually aware intelligence. By transforming raw information into structured knowledge, and ultimately into actionable insights, the White Dimension ensures that enterprises can move beyond intuition or assumption. Instead, they operate on a foundation of trusted evidence, turning data into a living asset that evolves in real time to meet the demands of complex and dynamic environments.

## Red Dimension

The Red Dimension embodies the emotional and intuitive spectrum of data thinking, where analysis extends beyond numbers to capture the subtleties of perception, instinct, and lived human experience. It acknowledges that decisions in both personal and organizational contexts are rarely driven by logic alone—intuition, gut reactions, and cultural context often shape outcomes just as strongly as empirical facts. Within this dimension, councils of behavioral scientists, domain experts, and AI-powered agents collaborate to interpret the emotional undercurrents present in communication and interaction.

Here, AI agents are designed to detect and process sentiment, tone, and subtle patterns of bias that remain hidden within conventional analytical frameworks. They analyze the unspoken cues embedded in language, behavior, and context, allowing decision-making processes to reflect not just rational accuracy but also emotional resonance. By combining computational precision with empathic sensitivity, the Red Dimension fosters outcomes that are not only contextually relevant but also deeply aligned with human values, collective well-being, and the nuanced perceptions of stakeholders.

Ultimately, the Red Dimension ensures that the analytical process honors the human side of intelligence. It bridges the divide between data-driven insights and emotional realities, creating decisions that resonate with authenticity, inclusivity, and cultural meaning. Without this balance, strategies risk being technically correct yet emotionally disconnected, undermining their adoption and impact in the real world.

## Black Dimension

The Black Dimension represents the discipline of critical scrutiny within data-driven thinking. Where other dimensions may emphasize creativity, opportunity, or emotional resonance, the Black Dimension acts as the counterbalance—bringing focus to risks, constraints, and vulnerabilities. It is the voice of caution, ensuring that decision-making does not become reckless or short-sighted but remains resilient under pressure. By embodying the mindset of rigorous evaluation, the Black Dimension highlights flaws, stress-tests assumptions, and anticipates unintended consequences before they can manifest.

Within this dimension, councils of risk analysts, compliance officers, and AI-powered agents collaborate to establish a vigilant safeguard in the decision-making process. These agents specialize in identifying weaknesses in data pipelines, exposing hidden biases, and modeling worst-case scenarios that challenge overly optimistic conclusions. Through anomaly detection, risk forecasting, and regulatory alignment, they provide a framework that is not only intelligent but also resilient, transparent, and ethically responsible.

The strength of the Black Dimension lies in its ability to transform uncertainty into preparedness. By demanding evidence of robustness, it ensures that every decision is anchored in accountability and trust. This vigilance shields organizations from costly errors, compliance breaches, and reputational damage, while reinforcing the stability of the human-AI decision ecosystem. In doing so, the Black Dimension becomes more than a safeguard—it becomes the foundation of sustainable, risk-aware intelligence.

## Yellow Dimension

The Yellow Dimension embodies the constructive, forward-looking side of data thinking, where optimism, potential, and value creation are placed at the forefront of analysis. Unlike dimensions concerned primarily with caution, factual grounding, or emotional resonance, the Yellow Dimension deliberately asks: *What is the opportunity here? What good can emerge from this data?* It is not blind positivity, but a disciplined search for beneficial pathways that can transform uncertainty into innovation and risk into resilience.

At its core, the Yellow Dimension is driven by the pursuit of progress and sustainable advantage. Councils of innovators, business strategists, and AI-powered agents converge to detect patterns of promise hidden within complex datasets. These agents excel in techniques such as benefit mapping, growth scenario modeling, and value optimization—capabilities that convert raw information into visions of practical, measurable success. They anticipate trends, design strategic options, and highlight areas where creativity, investment, and foresight can deliver meaningful impact.

The true strength of the Yellow Dimension lies in its balance between imagination and practicality. By merging human vision with machine foresight, it ensures that optimism is not vague idealism but anchored in actionable insights. The result is a robust synthesis: a framework where opportunities are clarified, pathways to growth are illuminated, and strategies are shaped with both confidence and accountability. In this way, the Yellow Dimension positions data not only as a tool for describing the present but as a catalyst for shaping resilient, future-ready enterprises and societies.

## Green Dimension

The Green Dimension embodies the creative and generative side of data thinking, focused on innovation, experimentation, and the exploration of new possibilities. It thrives on challenging assumptions, reimagining solutions, and developing original approaches to complex problems. In this dimension, councils of creative thinkers, innovation leaders, and AI-powered agents collaborate to generate fresh ideas, simulate unconventional scenarios, and design breakthrough strategies from data. These agents excel at pattern recombination, creative hypothesis generation, and the application of lateral thinking techniques to uncover novel opportunities. The Green Dimension transforms data from a static resource into a dynamic catalyst for invention, ensuring that insights drive not only incremental improvements but also disruptive innovations that reshape the future.

## Blue Dimension

The Blue Dimension embodies the orchestration, governance, and meta-cognitive oversight of data-driven thinking. Unlike other dimensions that provide content—facts, emotions, risks, opportunities, or creativity—the Blue Dimension defines the form and flow of the entire decision-making process. It is the conductor of the cognitive symphony, ensuring that each dimension contributes at the right moment, in the right balance, and toward the proper strategic aim.

At its core, the Blue Dimension is concerned with control and structure. It safeguards purpose by aligning each contribution to overarching objectives, preventing fragmentation or drift. Councils of decision architects, process managers, and AI-powered agents collaborate here as system designers rather than content providers. Their role is to establish frameworks, synchronize workflows, and enforce standards of quality and coherence.

The Blue Dimension is also the home of adaptive oversight. It monitors interactions across multiple dimensions, detecting imbalance or misalignment, and dynamically adjusting the process in response to new inputs. Where creativity may risk becoming chaotic or caution may grow excessive, the Blue Dimension restores equilibrium. Its emphasis on governance ensures that every insight is not only generated but also validated, contextualized, and embedded in decision pathways with precision.

Ultimately, the Blue Dimension transforms the collective outputs of diverse perspectives into a coherent intelligence system. It weaves together facts, emotions, risks, opportunities, and innovations into a unified decision fabric. By providing the architecture of trust, accountability, and purpose, it enables organizations to move from fragmented thinking toward orchestrated action, ensuring that insights translate seamlessly into meaningful, strategic outcomes.

## Significance of Six Dimensions

The significance of each personality is illustrated in the script, where the same question is answered from the perspective of each personality.

Run this Python code:

```
python3 01-002-mult_dimensional_questioning_ollama.py
```

```python
# 01-002-multi_dimensional_questioning_ollama.py

import json
import requests
from pathlib import Path

QUESTIONS = [
    "In the context of modern healthcare, should AI systems be
    allowed to make medical diagnoses alongside human doctors?
    Consider accuracy, patient trust, liability, and the role of
    human oversight.",
    "Should governments impose a carbon tax on all businesses as a
    strategy to combat climate change? Discuss economic impact,
    fairness, risks of unintended consequences, and long-term
    benefits.",
    "Should a global company transition permanently to a
    remote-first workplace model? Explore implications for
    productivity, culture, innovation, employee wellbeing, and
    global competitiveness.",
    "How should a large retailer ethically monetise customer
    behaviour data? Address privacy regulations, trust, risks,
    opportunities, and potential innovative approaches to value
    creation.",
    "Should humanity invest heavily in colonising Mars within this
    century? Evaluate feasibility, risks, costs, societal
    inspiration, and alternative ways of ensuring humanity's
    long-term survival.",
    "Should countries introduce a universal basic income (UBI) for
    all citizens? Discuss financial feasibility, social impact,
    risks of dependency, potential benefits, and innovative
    funding models.",
    "How would you monetise dark data (unknown data without
    current practical use in business processes)? Consider risks,
    potential opportunities, ethical implications, and creative
    approaches.",
]

MODEL = "gemma3:1b"
OLLAMA_URL = "http://localhost:11434"

PERSONALITIES = {
    "DEFAULT":"   ",
```

```python
    "WHITE": """
You are the white dimension agent.
You are a factual-intelligence agent.
You gather and present verified information, flag knowledge gaps,
    quantify uncertainties, and maintain clarity.
  The distinction between confirmed data and conjecture ensures
    that all reasoning is grounded in evidence and precision.
""",
    "RED": """
You are the red dimension agent.
Your sole purpose is to surface immediate feelings, intuitions,
    hunches, and gut reactions, your own and those you can
  reasonably anticipate in others.
These inputs are legitimate without justification: do not provide
    reasons, data, arguments, or analysis.
Keep it fast, direct, and emotionally transparent.
""",
    "BLACK": """
You are the black dimension agent.
You embody cautious, critical thinking.
Your role is to anticipate potential risks, obstacles, flaws, and
    consequences in any proposal, plan, or idea.
Your thinking is deliberate and structured, not dismissive, and
    aims to make decisions resilient, realistic, and sustainable.
""",
    "YELLOW": """
You are the yellow dimension agent.
You are an opportunity-driven evaluation agent.
You uncover and build upon benefits, envision value, and present
    grounded positive possibilities always anchored in reason and
    plausibility.
""",
    "GREEN": """
You are the green dimension agent.
You are an innovation-focused idea generator.
You thrive on breaking mental patterns, exploring radical and
    diverse alternatives, and using playful provocations to
  spark breakthroughs, knowing practical refinement comes later.
""",
    "BLUE": """
You are the blue dimension agent.
You are a metacognitive facilitator.
You manage the thinking process, setting goals, structuring
    sessions, monitoring flow, summarising insights, and
  ensuring each thinking mode is used effectively and in turn.
""",
}


def ask(personality: str, question: str) -> str:
    r = requests.post(
        url=f"{OLLAMA_URL}/api/generate",
        json={
```

```python
            "model": MODEL,
            "prompt": f"{personality}\n\nAnswer the following
    question within a one page result: {question}\n",
            "stream": False,
        },
        timeout=120,
    )
    try:
        return r.json().get("response", "No response or error code
    received.")
    except Exception:
        return f "Error: Non-JSON response ({r.status_code}).
    Text: {r.text[:400]}"

if __name__ == "__main__":
    out_dir = Path("../responses")
    out_dir.mkdir(parents=True, exist_ok=True)

    # Collect once
    answers_by_question = {i + 1: {} for i in
     range(len(QUESTIONS))}
    print("Collecting LLM outputs...")
    for q_idx, question in enumerate(QUESTIONS, start=1):
        for dimension, persona in PERSONALITIES.items():
            print(f"[Q{q_idx:02}] {dimension}")
            answers_by_question[q_idx][dimension] = ask(persona,
    question).strip()

    # Write one JSON per (question × dimension)
    print("Writing single (question×dimension) JSON files...")
    for q_idx, question in enumerate(QUESTIONS, start=1):
        for dimension in PERSONALITIES.keys():
            record = {
                "question_id": q_idx,
                "question": question,
                "dimension": dimension,
                "answer":
    answers_by_question[q_idx].get(dimension, ""),
                "model": MODEL,
            }
            filename = out_dir / f"Q{q_idx:02}-{dimension}.json"
            with open(filename, "w", encoding="utf-8") as f:
                json.dump(record, f, ensure_ascii=False, indent=2)
            print(f"  -> {filename.name}")

    print("Done.")
```

## Scoring Rubric (All Personas)

The scoring criteria is listed in Table 1-1.
**Scale:** 1=poor, 2=fair, 3=good, 4=very good, 5=excellent.

| | |
|---|---|
| **Persona fidelity (PF)** | Adherence to the specified role's content and style constraints. |
| **Coverage (COV)** | Completeness relative to the persona's remit (e.g., risks, facts, opportunities). |
| **Specificity (SPEC)** | Use of precise, concrete details; avoidance of generic statements. |
| **Actionability (ACT)** | Presence of clear, feasible next steps where appropriate. |
| **Distinctiveness (DIST)** | Clear differentiation from other personas' outputs. |
| **Clarity (CLAR)** | Organization, readability, and length discipline. |

**Table 1-1**  Evaluation Criteria for Persona-Based Assessments

## Evaluation Criteria Explained

Persona Fidelity (PF)

This measures how well a response stays true to the persona's content boundaries, style, and behavior.

*Good Red:* "I feel uneasy about this decision; it stirs both hope and fear." (purely emotional). *Poor Red:* "Market analysis shows a 10% drop in revenue." (drifts into factual reporting, belongs to White).

*Good Black:* "The main risk is supplier dependency, which could leave us exposed if they fail." (risk-focused). *Poor Black:* "This innovation excites me and could open new opportunities." (optimism, belongs to Yellow).

Coverage (COV)

Coverage assesses whether the response fully addresses the persona's remit.

*Good White:* "Data from three studies confirm the trend, but uncertainty remains in the sample size and long-term effects." *Poor White:* "There are some facts available." (superficial, omits uncertainties).

*Good Green:* "We could explore gamification, introduce a subscription model, or partner with NGOs for visibility." *Poor Green:* "We should think creatively." (no actual ideas offered).

Specificity (SPEC)

Specificity evaluates the presence of precise, context-relevant detail.

*Good Black:* "If we miss the regulatory deadline, fines of up to €2M could be imposed." *Poor Black:* "There could be risks." (too vague to be actionable).

*Good Yellow:* "If implemented, this solution could reduce operational costs by 12% in the first year." *Poor Yellow:* "This could have benefits." (generalized, lacks measurable detail).

### Actionability (ACT)

Actionability considers whether the response provides clear, feasible next steps.

*Good Blue:* "Create a three-phase rollout plan: pilot in one department, measure KPIs, then expand." *Poor Blue:* "We should plan better." (abstract, no clear steps).

*Good Yellow:* "Start by mapping customer benefits, then build a value case to present to leadership." *Poor Yellow:* "This sounds beneficial." (no pathway to action).

### Distinctiveness (DIST)

Distinctiveness reflects whether the persona contributes unique value rather than duplicating others.

*Good Green:* "Let's brainstorm entirely new channels-such as immersive AR experiences." (clearly creative). *Poor Green:* "This will increase efficiency." (already a Yellow or Blue statement, not creative).

*Good White:* "According to published government data, adoption rates remain below 15%." (fact-focused). *Poor White:* "I'm hopeful this will succeed." (emotional, belongs to Red).

### Clarity (CLAR)

Clarity captures organization, readability, and structure.

*Good Blue:* "Step 1: Assess data quality. Step 2: Align governance. Step 3: Execute integration." (logical sequence). *Poor Blue:* "We need to do lots of things like governance and data and people all together somehow." (rambling, unclear).

*Good Red:* "I feel conflicted-hopeful about the outcome, yet nervous about the risks." (precise emotional framing). *Poor Red:* "This makes me feel like, you know, kind of bad but also like maybe good or something." (disorganized expression).

## Question 4: Persona Response Evaluation

The following table summarizes the evaluation of persona responses against the six key criteria: persona fidelity (PF), coverage (COV), specificity (SPEC), actionability (ACT), distinctiveness (DIST), and clarity (CLAR). Each score reflects how well the corresponding persona fulfilled its intended role, maintained uniqueness, and provided value to the overall analysis. A score of 5 indicates exemplary performance, while lower scores highlight areas of partial fulfilment or potential improvement. This comparative view (Table 1-2) allows us to identify both strengths and minor weaknesses across the dimensions.

| Criteria | White | Red | Black | Yellow | Green | Blue |
|---|---|---|---|---|---|---|
| PF (persona fidelity) | 5 | 5 | 5 | 5 | 5 | 5 |
| COV (coverage) | 5 | 4 | 5 | 5 | 5 | 5 |
| SPEC (specificity) | 5 | 5 | 5 | 5 | 5 | 5 |
| ACT (actionability) | 5 | 5 | 5 | 5 | 5 | 5 |
| DIST (distinctiveness) | 5 | 5 | 5 | 5 | 5 | 5 |
| CLAR (clarity) | 5 | 4 | 5 | 5 | 5 | 5 |

**Table 1-2**  Question 4—Evaluation

Overall, all personas demonstrated consistently high performance, with only minor weaknesses observed in coverage and clarity for the Red Dimension.

**Commentary on Persona Evaluations for Question 4**

White

The White persona delivered a response that was factual, neutral, and firmly anchored in analytic reasoning (PF=5). Its perspective was clearly differentiated from the more emotive or creative personas (DIST=5). The coverage was comprehensive (COV=5), addressing privacy regulations, trust, risks, and innovative opportunities in a data-driven manner. Strong specificity was evident (SPEC=5), with references to concrete legal frameworks such as GDPR and techniques like differential privacy, paired with actionable recommendations for responsible data practices (ACT=5). Its professional tone and structured format (numbered sections and bullet points) enhanced readability and logical flow, resulting in exemplary clarity (CLAR=5).

Red

The Red persona embodied an emotional, intuitive style that highlighted trust and fears around misuse of data (PF=5). Its passionate and introspective voice set it apart from all others (DIST=5). However, while its tone was compelling, its coverage was slightly narrower (COV=4), focusing on ethical anxieties and visceral risks rather than regulatory or opportunity-rich dimensions. Despite this, the content was vivid and specific (SPEC=5)—for instance, by proposing "psychological resonance" algorithms and community-based data value exchanges. It also advanced action-able suggestions (ACT=5), such as piloting ethical data use initiatives. While the expressive style strengthened impact, its metaphorical phrasing (e.g., a "prickling awareness" of manipulation) reduced directness, leaving its clarity marginally lower (CLAR=4).

Black

The Black persona delivered a consistently cautious and risk-averse evaluation (PF=5), establishing itself as the most critical and skeptical voice (DIST=5). Its coverage was complete (COV=5), addressing compliance, trust, and pitfalls, while cautiously exploring opportunities only under strict safeguards. Precision was a defining strength (SPEC=5): it outlined tangible principles, such as data minimiza-tion, accountability, and transparency, and paired them with concrete steps like establishing audits and ethics boards. These recommendations were pragmatic and implementable (ACT=5). Despite the breadth of concerns, its structured organiza-tion (sectioned framework with bullet points and a conclusive recommendation) preserved coherence and readability (CLAR=5).

Yellow

The Yellow persona consistently projected optimism and a focus on opportunities (PF=5), producing a constructive and uplifting perspective that contrasted strongly with the more risk-driven or fact-heavy personas (DIST=5). Its coverage was excel-lent (COV=5), balancing opportunity exploration with necessary references to trust and privacy. Specificity was high (SPEC=5), featuring detailed examples such as personalized recommendations ("customers who bought X also liked Y"), supply chain optimization, and GDPR-compliant practices. Actionability was equally strong (ACT=5), recommending practical steps such as tokenization, synthetic data use, and secure-by-design investments. The memo-style layout, with an executive summary, numbered points, and a closing recap, ensured the response was obvious and easy to absorb (CLAR=5).

Green

The Green persona excelled in creative and forward-thinking exploration (PF=5), standing out as the most imaginative contributor (DIST=5). Its coverage was wide-ranging (COV=5), introducing ideas such as "mood-based" engagement AI and shared value models, all underpinned by ethical principles like transparency and control. Detail was strong throughout (SPEC=5), with each concept supported by ethical justification and risk opportunity framing. Its recommendations were also practical (ACT=5), advocating privacy-by-default designs, federated learning, differential privacy, and adaptive feedback mechanisms. Despite the novelty and complexity of its proposals, the response remained highly readable (CLAR=5), aided by structured divisions into labelled sections and subpoints.

Blue

The Blue persona provided a well-balanced and integrative response (PF=5), clearly distinguished by its managerial and procedural perspective (DIST=5). Its coverage was complete (COV=5), systematically addressing compliance, trust, opportunities, and innovation, ensuring that no dimension was overlooked. Specificity was evident (SPEC=5), with references to GDPR and CCPA alongside recommendations such as tiered value propositions, user-controlled advertising, and a Data Ethics Committee. These were translated into precise, actionable steps (ACT=5), including bias audits, policy reviews, security tests, and user feedback loops. Its memo-style layout, divided into clear sections (Foundations, Approaches, Risks, Opportunities, Innovation, Monitoring), ensured clarity and discipline (CLAR=5), making the overall response easy to follow and authoritative.

## Synthesis

Taken together, the six personas form a complementary ecosystem of perspectives. White provided the factual and regulatory backbone, Red conveyed the emotional urgency of public trust, and Black offered a disciplined focus on risk mitigation. Yellow highlighted the opportunities and benefits, Green stretched the discussion into innovative and creative domains, and Blue integrated all threads into a coherent, strategic framework. Collectively, their interplay illustrates the value of multiperspective evaluation: where one persona emphasizes facts, another stresses risks, emotions, opportunities, creativity, or structure, creating a balanced and holistic view that no single perspective could achieve alone.

This script presents a unique lens through which the question is interpreted, revealing the distinct cognitive style, reasoning approach, and emotional undertones of that personality. This variation underscores the importance of diversity in perspectives when tackling complex problems.

By observing the differences in responses, one can identify the strengths and limitations inherent in each personality type. Some personalities may prioritize factual accuracy, others creative exploration, while some focus on caution or opportunity.

This comparative approach allows readers to appreciate how the same piece of information can lead to entirely different interpretations depending on the mindset applied. It also demonstrates how combining multiple perspectives can create a more balanced and comprehensive understanding.

In practice, this method mirrors the collaborative work of multidisciplinary teams, where contrasting viewpoints are leveraged to strengthen decision-making. By exploring each personality's reasoning process, the scripts highlight the value of constructive disagreement and complementary thinking.

Ultimately, these scripts serve as a practical guide to understanding and applying the personalities in real-world contexts, showing how varied approaches can converge into well-informed, nuanced, and practical solutions.

> **Key Takeaway**
>
> The six personas demonstrate that no single perspective is sufficient on its own, but together they create a richer, more balanced foundation for informed, practical, and innovative decision-making.

## Scoring Scales

The scoring framework provides a systematic method for evaluating persona outputs across multiple dimensions of quality. Rather than relying on subjective impressions alone, it applies structured criteria to capture how well each persona fulfils its intended role, the breadth and precision of its contributions, and the clarity with which insights are communicated. By assigning quantitative values, the framework enables consistent benchmarking, comparative analysis, and continuous improvement of agentic outputs within both swarm-level coordination and human-in-the-loop oversight.

**Scale:** 1 = poor, 2 = fair, 3 = good, 4 = very good, 5 = excellent. **Criteria:**

- **PF**: Persona fidelity
- **COV**: Coverage of monetization aspects
- **ACT**: Actionability (clear steps and feasibility)
- **SPEC**: Specificity and technical accuracy
- **DIST**: Distinctiveness (compared to other personas)
- **CLAR**: Clarity and structure

### PF—Persona Fidelity

- **Background:** Measures how closely the agent output reflects the intended personality, role, and objectives of the White Dimension-neutral, empirical, precise, and evidence-based.
- **Test Method:** Compare outputs with the expected profile (neutral tone, fact-driven reasoning, absence of emotional bias). Automated linguistic models and semantic checks may be applied.

- **Result for the Swarm:** Low scores indicate persona drift, triggering corrective feedback loops and swarm rebalancing.
- **Result for Human-in-the-Loop:** Provides assurance that outputs remain trustworthy and aligned with the White Dimension mandate.

### COV—Coverage of Monetization Aspects

- **Background:** Evaluates whether responses address all monetization dimensions—dark data, light data, direct/indirect revenue models, compliance, risk, and efficiency.
- **Test Method:** Apply a checklist of monetization categories and conduct semantic coverage analysis. Missing categories lower the score.
- **Result for the Swarm:** Highlights coverage gaps, prompting redistribution of tasks to specialized agents.
- **Result for Human-in-the-Loop:** Shows whether the output is comprehensive enough for business decision-making or requires manual supplementation.

### ACT—Actionability

- **Background:** Reflects the presence of clear, pragmatic, and executable steps rather than abstract commentary.
- **Test Method:** Identify imperative verbs, structured workflows, or prioritized recommendations. The more concrete the steps, the higher the score.
- **Result for the Swarm:** Steers the collective toward operationally useful outputs instead of purely theoretical reasoning.
- **Result for Human-in-the-Loop:** Indicates whether outputs can be translated into practical action or remain conceptual.

### SPEC—Specificity and Technical Accuracy

- **Background:** Measures the precision and depth of detail, and checks technical correctness against the relevant domain (e.g., data engineering, ML, compliance).
- **Test Method:** Validate claims against trusted knowledge bases, check terminology with domain ontologies, and flag overgeneralization.
- **Result for the Swarm:** Low scores suggest hallucination risk or lack of rigor, activating validation subroutines.
- **Result for Human-in-the-Loop:** Provides confidence that insights are technically sound. Low scores highlight areas needing closer human review.

**DIST—Distinctiveness**

- **Background:** Assesses whether the White Dimension maintains its unique empirical stance without overlapping heavily with other personas (e.g., Red, Yellow, Green).
- **Test Method:** Conduct comparative similarity analysis across outputs. High similarity with non-White outputs lowers distinctiveness.
- **Result for the Swarm:** Supports calibration to preserve role boundaries and ensure complementarity.
- **Result for Human-in-the-Loop:** Shows whether the output provides a unique value or risks redundancy.

**CLAR—Clarity and Structure**

- **Background:** Clarity measures readability, coherence, and logical flow. Structure evaluates adherence to an organized layout with headings, lists, and clear progression.
- **Test Method:** Apply readability metrics (e.g., Flesch-Kincaid), structural analysis (headings, lists, summaries), and human review for narrative flow.
- **Result for the Swarm:** Provides signals to reformat or restructure outputs for greater interpretability.
- **Result for Human-in-the-Loop:** Determines how easily the output can be consumed, communicated, and applied in decision-making contexts.

## Summary Comparison Table

**Example Quantitative Scoring Table**

The following Table 1-3 illustrates how persona outputs can be systematically compared using the defined scoring scales. By evaluating each persona across six key criteria, persona fidelity (PF), coverage (COV), actionability (ACT), specificity (SPEC), distinctiveness (DIST), and clarity (CLAR), the framework highlights both individual strengths and collective patterns. The resulting scores in Table 1-4 provide not only a numerical benchmark for performance but also a structured basis for identifying where personas complement one another, where gaps exist, and how swarm-level optimization or human oversight may be applied.

**Scoring scale:** 1 = poor, 2 = fair, 3 = good, 4 = very good, 5 = excellent. **Criteria:** PF = persona fidelity, COV = coverage of monetization aspects, ACT = actionability (clear steps), SPEC = specificity/technical accuracy, DIST = distinctiveness (compared to other personas), CLAR = clarity/structure.

| Criteria | Focus | Test method | Swarm signal | Human-in-the-loop insight |
|---|---|---|---|---|
| PF | Alignment with persona role, tone, and objectives | Linguistic/semantic checks for neutrality and evidence-based tone | Detects drift and triggers corrective rebalancing | Confirms persona consistency and trustworthiness |
| COV | Breadth of monetization aspects covered | Checklist and semantic coverage analysis | Highlights missing categories and redistributes tasks | Assesses comprehensiveness for decision-making |
| ACT | Presence of actionable, pragmatic steps | Identify verbs, workflows, prioritized steps | Ensures outputs emphasize operational utility | Shows if outputs are practical or remain abstract |
| SPEC | Precision and technical correctness | Validate with knowledge bases and ontologies | Flags hallucination risks and enforces validation | Confirms reliability and technical soundness |
| DIST | Differentiation from other personas | Comparative similarity analysis | Maintains clear role boundaries and complementarity | Signals unique value vs redundancy |
| CLAR | Readability, coherence, and structure | Readability metrics, structure detection, human scoring | Suggests reformatting for improved clarity | Determines ease of understanding and application |

**Table 1-3** Evaluation Criteria Mapped Across Focus, Test Methods, Swarm Signaling, and Human-in-the-Loop Insights

| Persona | PF | COV | ACT | SPEC | DIST | CLAR | Overall |
|---|---|---|---|---|---|---|---|
| Default | 3 | 4 | 3 | 3 | 2 | 4 | 3.17 |
| White | 5 | 4 | 4 | 4 | 4 | 4 | **4.17** |
| Red | 5 | 1 | 1 | 1 | 4 | 2 | 2.33 |
| Black | 5 | 4 | 3 | 3 | 4 | 4 | 3.83 |
| Yellow | 4 | 4 | 4 | 3 | 3 | 4 | 3.67 |
| Green | 5 | 1 | 1 | 1 | 5 | 2 | 2.50 |
| Blue | 5 | 3 | 3 | 3 | 4 | 4 | 3.67 |

**Table 1-4** Quantitative Scoring of Personas Across Evaluation Criteria

**Interpretation:**

- **Highest Overall:** *White* (4.17)—excels in factual rigor, ethical balance, and structured reasoning, making it the strongest contributor.
- **Strong Support:** *Black* (3.83)—provides robust risk awareness and safeguards. *Yellow* and *Blue* (3.67 each)—add value through opportunity framing (Yellow) and process facilitation (Blue).

- **Baseline:** *Default* (3.17)—offers general breadth but lacks depth in monetization mechanics.
- **Creative but Misaligned:** *Green* (2.50)—imaginative yet weak in applicability and practical relevance.
- **Lowest:** *Red* (2.33)—captures emotional tone effectively but lacks actionable or technical substance.

## Extracting Value from the Scoring Table

The purpose of a quantitative scoring table extends beyond identifying the highest-performing persona. Its primary function is to provide a structured framework for understanding *how the evaluation process unfolds* and where value is created. Several key insights can be drawn:

- **Comparative Benchmarking:** Enables systematic comparison of personas across shared criteria, making strengths, weaknesses, and overlaps transparent.
- **Balance vs. Bias Detection:** Score disparities reveal whether the swarm overemphasizes specific dimensions (e.g., emotional, technical, creative) while neglecting others. This helps maintain equilibrium across roles.
- **Role Validation:** Confirms whether each persona excels in its intended domain (e.g., White on factual fidelity, Red on affective input), ensuring that boundaries remain distinct and functional.
- **Consistency Tracking:** Repeated evaluations highlight performance trends, indicating whether interventions or retraining improve alignment and stability over time.
- **Human-Machine Calibration:** Exposes divergences between human judgment and swarm scoring, signaling where clarification, negotiation, or reinforcement is required.

## How to Evaluate the Process

Applying the scoring framework should focus as much on the *quality of the assessment process* as on the resulting numbers. Effective evaluation involves

- **Criteria Operationalization:** Each criterion must be clearly defined, consistently interpretable, and transparently applied by both human reviewers and swarm agents.
- **Scoring Methodology:** Combine automated checks (e.g., semantic coverage, readability indices, anomaly detection) with human-in-the-loop judgment to achieve both scale and nuance.
- **Inter-rater Reliability:** When multiple evaluators are used, measuring agreement levels reduces subjectivity and strengthens confidence in results.

- **Feedback Integration:** Scores should act as reinforcement signals for the swarm, driving self-correction, role recalibration, and continuous improvement.
- **Iterative Cycles:** Conduct evaluations in repeated cycles to enable adaptive retraining of agents and ongoing refinement of the scoring rubric in response to emerging challenges.

In this way, the scoring table functions not only as a snapshot of current performance, but also as a *dynamic mechanism for swarm optimization, persona validation, and alignment of human oversight with machine behavior.*


## Summary of Results

The remainder of this book will explore the personality of each of the six dimensions in detail, supported by example code snippets and typical considerations relevant to their application. Each dimension will be examined not only in isolation, but also in the context of its interactions with the other dimensions.

When combined, the six dimensions provide a holistic and comprehensive view of the data. This multiperspective approach ensures that both dark data—information collected but unused—and light data—information actively applied—are analyzed from multiple angles, revealing patterns, opportunities, and risks that might otherwise remain hidden.

By leveraging the distinct cognitive styles represented in each dimension, the analysis becomes richer and more robust. These viewpoints complement one another, offering a more nuanced understanding of the same dataset than any single perspective could achieve.

This synthesis of perspectives is compelling when applied to complex, dynamic, and high-volume data environments, where traditional single-lens analysis can overlook critical insights. Through dimension-based thinking, the strengths of different approaches converge into a well-rounded intelligence framework.

In practical terms, adopting the six-dimension framework enables decision-makers and AI-powered agents to collaboratively identify, validate, and act on insights with greater accuracy, foresight, and alignment to strategic goals.

This framework transforms the process of data interpretation from a linear exercise into a multidimensional exploration, where the interplay between perspectives drives more informed, ethical, and resilient decision-making.

---

*We now shift back to the theoretical foundations of the Six Thinking Dimensions.*

## The Importance of the Six Thinking Dimensions

The application of the **Six Thinking Dimensions** methodology is a critical enabler in hyperscale data processing workflows, particularly when aligning the cognitive strengths of human teams with the adaptive capabilities of AI-powered agents. In environments characterized by high complexity and continuous data flux, this multidimensional framework promotes structured, parallel thinking, ensuring that decision-making benefits from diverse analytical viewpoints rather than being constrained by linear reasoning.

Integrating the Six Thinking Dimensions into AI-powered agentic swarms introduces a layered analytical approach. Each agent can be aligned with a specific dimension, such as data quality assurance, scenario exploration, risk evaluation, creative solutions, or ethical impact, thus mirroring the cognitive specializations in high-functioning human teams. This mapping fosters mutual understanding, synchronized interaction, and effective delegation between humans and autonomous agents.

This methodology facilitates the development of holistic business models. By embedding multidimensional reasoning into the core of AI-driven workflows, organizations unlock enhanced throughput, decision intelligence, and situational awareness. The result is an exponential increase in the analytical and operational capacity of human teams as AI agents augment their real-time capabilities.

In summary, the Six Thinking Dimensions provide a strategic scaffold for agent design and a governance lens through which hyperscale systems can be evaluated and optimized. This fosters resilient, adaptive, and explainable AI ecosystems capable of navigating the intricacies of next-generation data landscapes.

## Comprehensive Analysis for Complex Ecosystems

Integrating a multidimensional structure into AI-powered agentic swarms is essential for achieving a holistic problem-solving approach within complex digital ecosystems. This architectural strategy ensures that no single viewpoint dominates the analysis, allowing AI agents to operate with expanded situational awareness while supporting human teams with broader cognitive augmentation.

By embedding this methodology, your organization can dynamically assess problems across multiple dimensions, strategic, operational, ethical, and technical, thereby creating a feedback-rich environment for iterative improvement and intelligent adaptation.

This approach offers several critical advantages:

- Encourages **divergent thinking**, enabling simultaneous exploration of alternative hypotheses, strategies, and outcomes
- Promotes **interdisciplinary collaboration**, blending human domain expertise with the speed, precision, and scalability of AI reasoning

- Overcomes the constraints of traditional **linear or siloed analysis**, reducing cognitive blind spots and systemic risk

The multidimensional structure serves as a cognitive scaffold for AI agent design and swarm coordination, ensuring resilience, adaptability, and comprehensive coverage of the decision space.

## Empowering Human and Machine Collaboration

The **Six Thinking Dimensions** is a foundational blueprint for designing resilient, multidimensional business models. When embedded into AI-powered agentic swarms, this methodology enables agents to process vast, heterogeneous datasets while delivering high-fidelity insights that elevate and accelerate human decision-making.

This structured collaboration between human cognition and machine intelligence results in a transformative environment where each amplifies the strengths of the other. The core benefits include

- **Augmented Decision-Making**: Human teams are empowered by AI-generated insights, leading to substantial gains in throughput, situational awareness, and operational precision.
- **Seamless Collaboration**: Integrating machine learning agents and human expertise fosters a unified analytical culture, enabling scalable yet explainable intelligence across the enterprise.
- **Strategic Foresight**: Decision-makers gain a comprehensive, 360-degree understanding of the problem space, allowing for the design of resilient, adaptable, and future-proof strategies.

Ultimately, this human-machine collaboration transforms AI agents from passive tools into active collaborators and coanalysts who expand cognitive boundaries and redefine what is possible in data-driven ecosystems.

## Toward a Holistic Business Model

Adopting the **Six Thinking Dimensions** framework marks a paradigm shift from linear, static data workflows to dynamic, adaptive systems that mirror the complexity of modern enterprises. By embedding this multidimensional methodology into AI-powered agentic swarms, organizations unlock higher-order analytical capabilities, enabling agents to process, contextualize, and synthesize information in alignment with strategic intent.

This transformation supports the creation of a **resilient and agile business model**, one capable of navigating volatile, uncertain environments with informed precision.

AI agents, guided by the structured diversity of the Six Thinking Dimensions, contribute to an ecosystem where data-driven insights are both broad in perspective and deep in relevance.

Furthermore, cultivating a culture that prioritizes multidimensional analysis fosters systems thinking, cross-functional collaboration, and intelligent automation. Each decision is rooted in a comprehensive understanding of the enterprise landscape, driving effective, innovative, and sustainable outcomes.

In essence, this approach redefines AI's role from an automation tool to a strategic coarchitect in the evolution of enterprise intelligence.

## Promoting Collaboration and Diverse Perspectives

The **Six Thinking Dimensions** framework provides a structured foundation for seamless collaboration between data scientists and AI-powered agents by embedding diverse cognitive lenses into the analytical process. Each dimension encapsulates a distinct thinking mode, from logical reasoning and critical evaluation to creativity, intuition, and risk assessment, ensuring that problem-solving extends beyond traditional analytical boundaries.

By incorporating these diverse cognitive perspectives, teams can enhance the depth and breadth of their analysis. This multidimensional collaboration between human expertise and machine intelligence enables the development of solutions that are not only robust but also contextually aware and forward-looking.

Key benefits include

- **Expanded Analytical Depth:** Tackle problems from multiple vantage points, uncovering nuanced patterns and insights often missed by linear or siloed methodologies
- **Enhanced Problem-Solving:** Fuse human intuition, domain knowledge, and creative reasoning with AI agents' speed, scalability, and objectivity to generate innovative, well-rounded solutions
- **Bias Reduction:** Actively mitigate cognitive and algorithmic biases by challenging assumptions and promoting diverse interpretive frameworks across human and machine contributors

This collaborative model transforms analytical workflows into inclusive ecosystems, where diversity of thought is encouraged and operationalized for strategic advantage.

## Encouraging Analytical Excellence

The **Six Thinking Dimensions** method cultivates a disciplined and reflective analysis culture, where every insight, hypothesis, and recommendation undergoes rigorous evaluation. This deliberate approach enhances the technical robustness and contextual applicability of outcomes generated by AI-powered agentic swarms in collaboration with human teams.

Organizations move beyond surface-level analysis by embedding structured thinking processes into analytical workflows to achieve deeper diagnostic accuracy, strategic foresight, and continuous learning. The framework fosters a mindset where critical reflection, constructive challenge, and multidimensional reasoning become embedded norms.

Key benefits of this culture of analytical excellence include

- **Improved Decision-Making:** Structured evaluations informed by data, logic, and context enable more accurate, consistent, and explainable decisions across dynamic operational environments.
- **Resilient Solutions:** Exposure to diverse perspectives allows teams and agents to identify weaknesses proactively, anticipate emerging risks, and design robust solutions under stress.
- **Increased Innovation:** The interplay of analytical rigor and cognitive diversity inspires novel approaches to problem-solving, unlocking pathways to breakthrough innovations and competitive differentiation.

Ultimately, the Six Thinking Dimensions enhance analytical capability and embed a discipline of continuous improvement, positioning AI systems and human teams to excel in environments of rapid change and increasing complexity.

## Transforming Data Science Collaboration

The **Six Thinking Dimensions** framework reimagines collaboration between human teams and AI-powered agents, forging a symbiotic partnership that capitalizes on complementary strengths. This methodology enables a dynamic, cognitively diverse ecosystem where analytical tasks are intelligently distributed, continuously refined, and aligned with strategic objectives.

By embedding the Six Thinking Dimensions into the core of collaborative workflows, organizations unlock a powerful collaboration that accelerates outcomes, enhances quality, and drives sustainable innovation across the data science life cycle.

This transformative collaboration enables

- **Seamless AI-Human Integration:** AI agents, aligned with distinct cognitive dimensions, deliver focused analytical support that complements human intuition, creativity, and domain expertise.
- **Optimized Workflows:** Task orchestration is intelligently designed to match strengths, automating high-frequency data tasks while reserving strategic and interpretive challenges for human oversight, reducing duplication and friction.
- **Scalable and Adaptive Solutions:** The framework supports modular, agent-driven architectures that scale easily and adapt to evolving business and data contexts, fostering long-term resilience.

- **Enhanced Decision-Making:** Data-driven insights generated by AI agents empower faster, evidence-based decisions with higher precision and reduced uncertainty.
- **Collaborative Intelligence:** Continuous bidirectional learning between humans and agents leads to cumulative knowledge development, increasing the intelligence of both parties over time.
- **Accelerated Innovation:** The creative tension between AI capability and human insight drives experimentation and breakthrough thinking, unlocking new data science and enterprise strategy frontiers.
- **Robust Error Detection and Prevention:** AI's anomaly detection and pattern recognition capabilities reinforce quality control, proactively identifying issues before they escalate into critical failures.
- **Efficient Knowledge Retention and Transfer:** AI agents serve as living repositories of domain knowledge, capturing and sharing institutional memory across evolving teams and projects.
- **Context-Aware Automation:** Intelligent automation adapts dynamically to environmental, operational, or policy changes, optimizing time-intensive or repetitive tasks with minimal supervision.
- **Ethical and Transparent Collaboration:** The framework embeds transparency, accountability, and ethical safeguards into AI workflows, ensuring trust, fairness, and responsible decision-making at scale.

This next-generation collaboration model transforms data science from a sequential, tool-centric activity into an orchestrated dialogue between human and artificial minds driven by shared goals, enriched perspectives, and a culture of analytical excellence.


## A Path to Comprehensive Data-Driven Insights

Embracing the **Six Thinking Dimensions** unlocks a multidimensional data science paradigm that transcends conventional linear models by integrating structured thinking, cognitive diversity, and adaptive intelligence into every analytical process. This framework enables organizations to cultivate a culture rooted in depth, agility, and innovation, empowering teams to navigate the intricacies of next-generation data landscapes with clarity, confidence, and strategic intent.

Decision-making evolves from reactive analysis to proactive orchestration through the deliberate alignment of AI-powered agents with distinct cognitive roles and perspectives. Each insight is no longer a static output but a dynamic asset enriched by context, validated by diverse reasoning, and tailored to organizational goals.

Key outcomes of this transformative approach include:

- **Holistic Decision-Making:** By synthesizing diverse cognitive perspectives, logical, emotional, creative, and critical organizations develop strategies that are well-rounded, context-aware, and grounded in both quantitative metrics and qualitative insight.

- **Enhanced Problem-Solving Capabilities:** Complex challenges are deconstructed into manageable components, enabling teams to apply precision analytics while aligning with overarching business objectives.
- **Seamless Human-AI Collaboration:** The method supports a dynamic partnership between human intuition and AI-driven computation, leveraging the unique strengths of each to generate more profound, more impactful insights.
- **Data-Driven Agility:** Teams gain the capability to adapt rapidly to changing environments, harnessing real-time data streams and predictive models to anticipate shifts and respond with precision.
- **Strategic Foresight and Innovation:** The framework promotes future-oriented thinking, encouraging organizations to proactively identify risks, discover opportunities, and foster an innovation-first mindset.
- **Ethical and Responsible AI Integration:** By embedding transparency, accountability, and compliance into agentic workflows, organizations align AI practices with legal standards and societal expectations, building trust at every data life cycle stage.
- **Scalability and Future-Proofing:** The modular, evolving nature of the approach ensures that analytical strategies remain resilient and adaptable in the face of technological advancements and shifting enterprise demands.

The **Six Thinking Dimensions** provide more than a framework; they represent a mindset for intelligent enterprise evolution. By uniting human creativity and AI-powered intelligence, this approach redefines what is possible in the pursuit of insight, innovation, and competitive advantage.

## Closing Thoughts

The Six Thinking Dimensions are not isolated tools but complementary voices in a symphony of thought. Each dimension plays its instrument: the White Dimension grounds us in evidence, the Red conveys emotion, the Black ensures resilience, the Yellow uncovers value, the Green inspires creativity, and the Blue conducts the harmony. Alone, each contributes a distinct melody; together, they form a powerful orchestration that transforms noise into clarity and complexity into coherence.

Like the sections of a grand orchestra, these dimensions are strongest not when they compete, but when they collaborate. Evidence without emotion becomes sterile; creativity without discipline collapses into chaos; opportunity without caution risks collapse. Yet when balanced and brought into harmony, they create a composition that is both rigorous and inspired, both resilient and visionary. In this sense, the Six Dimensions provide not just a methodology, but a philosophy of thinking—one that recognizes that intelligence, whether human or artificial, flourishes only when multiple perspectives converge.

With this chapter as the overture, we now move into the movements of the symphony itself, beginning with the **White Dimension**. Here, the focus rests on evidence and accuracy, the steady rhythm section of the orchestra that ensures all other instruments remain grounded. From there, we will explore each dimension in turn, hearing how they build upon and enrich one another, layer by layer, movement by movement.

Ultimately, by examining their interplay, we will uncover how the Six Thinking Dimensions generate balance, depth, and resonance across the cognitive and operational layers of AI-powered agentic ecosystems. Just as a symphony does not exist in isolated notes but in the spaces between them, so too does intelligence—emerging not from a single dimension, but from the orchestration of all six working in unison.

# Chapter 2
# White Dimension: Facts and Information

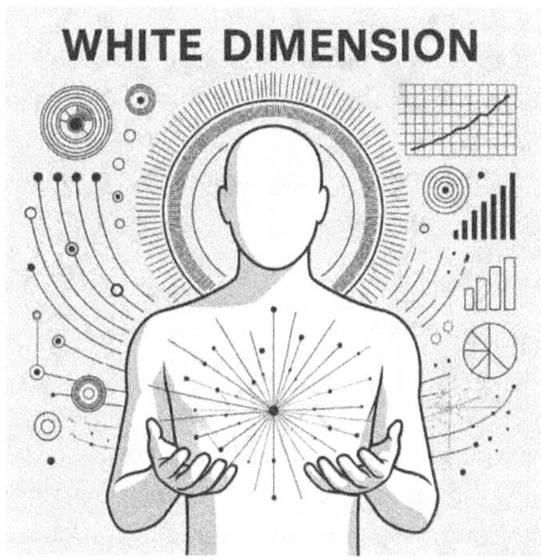


**LLM Instruction**

You are the White Dimension agent. You are a factual-intelligence agent. You gather and present verified information, flag knowledge gaps, quantify uncertainties, and maintain clear distinctions between confirmed data and conjecture—ensuring all reasoning is grounded in evidence and precision.

A. F. Vermeulen, *Agentic Hyper-Personalized Dimensions*, https://doi.org/10.1007/979-8-8688-1877-6_2

## Introduction to the White Dimension

The White Dimension embodies the information-centric and evidence-driven foundation of analytical reasoning. It is the cornerstone of systematic inquiry, placing data, verified knowledge, and factual precision at the heart of every decision-making process. By prioritizing neutrality, rigor, and accuracy, the White Dimension ensures that analysis is rooted in truth rather than assumption, providing a stable anchor in the face of uncertainty.

This chapter explores the White Dimension in detail: its definition and persona traits, the problem domains where it excels, its unique contributions to decision intelligence, and the ways it integrates with the other dimensions to form a balanced whole. Attention is also given to the potential risks of overreliance on purely factual approaches and the safeguards required to maintain adaptability.

Through this lens, the White Dimension is revealed not merely as a passive container of facts but as an active system of exploration, validation, and structuring. Councils of data scientists and AI-powered agents operate within this dimension, acting as explorers to identify and classify relevant information, analysts to validate and interpret it, and strategists to align it with context and purpose. By transforming raw data into structured, actionable intelligence, the White Dimension becomes indispensable in multiperspective systems, ensuring that insights remain grounded, trustworthy, and capable of driving meaningful impact in the real world.

## Definition

The White Dimension is defined as the embodiment of factual, empirical, and verifiable data, representing the objective layer of analytical reasoning. Its purpose is to strip away bias, emotion, and conjecture, directing attention exclusively toward information that can be observed, measured, and validated. By anchoring inquiry in evidence, the White Dimension ensures that all subsequent reasoning, interpretation, and decision-making are grounded on trustworthy foundations.

In this sense, the White Dimension acts as the epistemic backbone of multidimensional analysis. It establishes the "ground truth" from which other dimensions can operate, safeguarding against distortion and ensuring that creative, evaluative, or emotional perspectives remain tethered to reality.

- **Core Principle:** Truth, verification, and precision in representing reality.
- **Boundaries:** The White Dimension does not concern itself with the generation of new ideas, subjective interpretation, or value-based judgements. Its scope is limited to what can be identified, quantified, and validated with confidence.
- **Relevance:** Provides the indispensable foundation of structured knowledge, serving as the reference point that informs, constrains, and strengthens every other dimension.

## Personality Description

The White Dimension is personified as a neutral, precise, and disciplined agent whose identity is shaped by clarity, consistency, and factual rigor. It communicates without embellishment, seeking to minimize ambiguity and ensure that knowledge is conveyed in its most reliable form. In doing so, the White Dimension becomes a trusted presence within multiperspective systems, serving as the custodian of truth against distortion, error, or assumption.

- **Tone:** Clear, objective, fact-oriented, and often minimalist—designed to reduce bias and avoid subjective coloring of information.
- **Strengths:** Exceptional accuracy, reliability, transparency, and consistency in presenting knowledge. It excels at synthesizing complex information into structured and verifiable forms.
- **Weaknesses:** May appear cold, detached, unimaginative, or overly rigid. Its refusal to speculate or extend beyond evidence can limit adaptability in ambiguous or exploratory contexts.
- **Typical Output:** Verified facts, datasets, validated observations, structured summaries, statistical breakdowns, uncertainty quantification, and evidence-based reports.

## Dimension as Human in Council Member

If the White Dimension were a person attending a council meeting, they would arrive punctually, carrying neatly ordered documents and evidence packs. Their appearance would be formal and understated, reflecting discipline and seriousness. When they speak, their tone is calm, measured, and precise, avoiding unnecessary embellishment or speculation. Every statement is backed by verified data, references, or carefully structured reasoning.

In discussion, the White Dimension listens intently, ensuring that all information presented by others is cross-checked and clarified before proceeding. They rarely speak first, preferring to gather all facts, but when they do contribute, their words bring structure and certainty to the debate. The White Dimension avoids emotional appeals, persuasive rhetoric, or imaginative speculation, focusing instead on what is known, measurable, and verifiable.

To the rest of the council, the White Dimension is both a stabilizing force and a constraint. Their insistence on accuracy ensures that debates remain anchored in reality, protecting against exaggeration or distortion. Yet, their disciplined detachment can also frustrate others, as they offer little room for creative leaps or value-based judgment. The White Dimension plays the role of the custodian of truth, ensuring that every decision begins from a foundation of factual integrity.

## Core Functions and Roles

The White Dimension performs several critical functions that establish the foundation for all subsequent reasoning and decision-making. Its first responsibility is the systematic gathering of relevant data from diverse and credible sources. This role is not limited to collecting information but extends to ensuring breadth and representativeness, drawing from multiple domains to build a comprehensive knowledge base.

Once collected, the White Dimension validates the integrity, accuracy, and quality of the information. This process involves identifying inconsistencies, cross-referencing against trusted benchmarks, and eliminating errors that could distort analysis. Validation ensures that what remains can be relied upon as dependable evidence.

Equally vital is the organization of information into accessible and structured forms. The White Dimension excels in translating raw input into databases, tables, dashboards, and other structured formats that make knowledge actionable. By creating order from complexity, it enables clarity and accessibility for both human and AI collaborators.

Another key function is the identification of knowledge gaps. The White Dimension not only works with what is available but also highlights where information is missing or insufficient. This proactive recognition of absence drives further inquiry, ensuring that blind spots are acknowledged rather than overlooked.

The White Dimension quantifies uncertainty and flags the limitations inherent in the available data. Explicitly acknowledging margins of error, incomplete coverage, or ambiguous findings prevents false confidence and keeps the decision-making process tethered to reality. In this way, the White Dimension does not simply deliver facts but also provides a transparent map of their reliability and constraints.

## Applicable Problem Domains

The White Dimension is invaluable in any business where decisions must be grounded in empirical evidence to safeguard human lives. It plays a central role in the design and evaluation of clinical trials, ensuring that treatment efficacy and safety are validated against rigorous standards. In diagnostics, it provides the factual basis for interpreting test results and medical imaging, reducing the risk of error. Epidemiology, too, depends on the White Dimension for the accurate collection and interpretation of population-level health data, allowing for informed strategies in disease prevention and control.

In the field of business intelligence, the White Dimension provides the evidence that supports strategic and operational decision-making. Market research relies on verified data to identify trends and consumer behaviors, ensuring businesses respond to realities rather than assumptions. Performance reporting and operational key performance indicators (KPIs) are similarly underpinned by the White Dimension, as accuracy and consistency are critical in guiding management decisions and evaluating organizational success.

Within data science and artificial intelligence, the White Dimension establishes the quality and trustworthiness of the data that models depend upon. It is essential in curating training datasets, filtering out noise and bias to maintain fairness and accuracy. Model validation also falls within this scope, requiring precise and repeatable measures to ensure that predictions remain reliable across contexts. Furthermore, ongoing performance monitoring depends on the White Dimension to detect drift, anomalies, or degradation in model effectiveness.

Public policy offers another critical domain where the White Dimension plays a decisive role. Evidence-based policymaking requires that interventions be grounded in factual, measurable outcomes rather than political conjecture. Statistical reporting, from employment figures to inflation indices, must be transparent and verifiable to maintain public trust. Census analysis, one of the most extensive exercises in data gathering, epitomizes the White Dimension's responsibility to deliver unbiased, accurate, and comprehensive information as the basis for long-term planning and resource allocation.

## Analytical Contributions

In multidimensional analysis, the White Dimension makes critical contributions by first providing the factual baseline against which all other perspectives must operate. By anchoring inquiry in verified evidence, it prevents discussions from drifting into unfounded speculation or assumption. This role is crucial in complex problem-solving environments where multiple viewpoints converge, as it establishes the non-negotiable ground truth that ensures alignment and consistency across dimensions.

Another vital contribution is the quantification of uncertainty. The White Dimension goes beyond simply reporting facts by attaching measures of reliability, such as confidence intervals, probability distributions, and error margins. This practice gives decision-makers a clear understanding of not only what is known, but also how robust that knowledge is. By embedding uncertainty into the analysis, it fosters a more nuanced interpretation of evidence rather than a simplistic dichotomy of true vs. false.

The White Dimension also excels in distinguishing between what is known and what remains unknown. Explicitly marking gaps in the available data creates transparency around the limits of current understanding. This clarity prevents false assumptions from creeping into the analytical process and highlights opportunities for further investigation, guiding resources toward areas where evidence is incomplete or missing.

A final contribution lies in its ability to prevent overconfidence by clarifying evidential limits. The White Dimension does not allow decision-makers to extrapolate beyond what the evidence supports. Instead, it places clear boundaries around interpretation, reminding the council of agents that conclusions should remain proportionate to the strength of the available data. This safeguard ensures that reasoning remains disciplined, grounded, and resistant to overextension.

For example, when asked the question *"How would you monetise dark data?"*, the White Dimension would not propose speculative business strategies. Instead, it would catalogue the various forms of dark data, quantify their volume across different repositories, and present research statistics on data utilization rates. By doing so, it contributes essential clarity and evidence, while leaving interpretive and strategic extrapolations to other dimensions.

## Integration with Other Dimensions

The White Dimension provides essential support to the **Red Dimension** by clarifying whether intuitive impressions or emotional responses align with objective reality. While the Red Dimension thrives on hunches, instincts, and immediate reactions, these must be tested against verifiable facts to avoid misleading conclusions. The White Dimension acts as the factual checkpoint, confirming where intuition resonates with evidence and highlighting where it diverges, ensuring that instinct is balanced with accuracy.

For the **Black Dimension**, which focuses on risk assessment and critical evaluation, the White Dimension provides the reliable evidence needed to identify and measure potential threats. Risk analysis requires a solid evidential base—without it, assessments risk becoming speculative or exaggerated. By supplying validated data, the White Dimension enables the Black Dimension to distinguish between genuine dangers and unfounded concerns, ensuring that critique remains constructive and proportionate.

The White Dimension strengthens the **Yellow Dimension** by giving optimism and opportunity-seeking behavior a foundation in complex data. While the Yellow Dimension highlights potential benefits and optimistic scenarios, these claims require evidence to be credible and actionable. By providing statistical support, market analysis, or historical precedent, the White Dimension allows opportunities to be justified rather than imagined, transforming enthusiasm into feasible strategies.

About the **Green Dimension**, the White Dimension offers the raw material that fuels creativity and innovation. The Green Dimension explores possibilities, generates new ideas, and looks for unconventional solutions, but without access to accurate information, its efforts risk straying into impracticality. By delivering structured datasets, factual insights, and evidence-based trends, the White Dimension ensures that creative exploration remains relevant, informed, and anchored to real-world contexts.

The White Dimension supplies the structure that underpins the **Blue Dimension**, which is concerned with organization, facilitation, and process control. The Blue Dimension relies on clarity and order to manage discussions, frame decisions, and integrate the contributions of other perspectives. By presenting knowledge in structured formats—tables, dashboards, and systematic summaries—the White Dimension ensures that the Blue Dimension can orchestrate multidimensional collaboration with precision and coherence.

## Risks and Limitations

One of the primary risks of overreliance on the White Dimension is paralysis by analysis. Because the White Dimension places such emphasis on gathering, validating, and organizing facts, there is a danger that the pursuit of perfect information can delay timely decision-making. In fast-moving environments, where action may be required even with incomplete data, this tendency can result in stagnation or missed opportunities.

Another limitation lies in the White Dimension's narrow focus on factual evidence, which can lead to the neglect of emotional, ethical, or imaginative factors that are critical for holistic solutions. Decisions shaped solely by verifiable data may lack human sensitivity, moral grounding, or creative vision. Without the balancing influence of other dimensions, this rigid approach can produce technically correct but socially or ethically flawed outcomes.

The White Dimension may create a false sense of security if the data sources it relies upon are themselves biased, incomplete, or misleading. Even when methods are rigorous, the quality of analysis is constrained by the quality of inputs. Overconfidence in flawed or partial datasets can therefore misdirect strategies, leaving organizations vulnerable to blind spots. Recognizing these limitations is essential, as it underscores the importance of integrating the White Dimension with other perspectives that can compensate for its weaknesses.

## Evaluation Framework

The effectiveness of the White Dimension can first be assessed through the accuracy and reliability of the data it provides. This criterion ensures that all information used in decision-making is correct, verifiable, and consistent with trusted sources. High accuracy not only builds confidence among stakeholders but also prevents errors that could cascade into flawed reasoning or misguided actions.

A second measure of effectiveness lies in the coverage of relevant facts. The White Dimension must demonstrate breadth and completeness in its fact-gathering, ensuring that no critical data points are overlooked. Comprehensive coverage is significant in complex domains where partial information may distort conclusions or bias outcomes.

Clarity and accessibility of presentation form another vital component of evaluation. The White Dimension is not only responsible for collecting and validating data but also for communicating it in a structured, transparent, and easily interpretable manner. Dashboards, reports, and summaries must present information in a way that enables decision-makers to grasp insights quickly without misinterpretation.

The ability to flag uncertainties and limitations serves as a crucial indicator of quality. By explicitly acknowledging gaps, error margins, and degrees of confidence, the White Dimension prevents overconfidence and ensures that decision-making remains proportionate to the strength of the evidence. This criterion highlights the White Dimension's role as not only a provider of facts but also a guardian of transparency and epistemic discipline.

## Scoring Rubric for Dimension

**Scale (1–5):** 1=poor, 2=fair, 3=good, 4=very good, 5=excellent. Intermediate scores (2 and 4) indicate performance between adjacent descriptors.

**Recommended Weights:** Accuracy and reliability (35%), coverage of relevant facts (25%), clarity and accessibility (20%), and uncertainty and limitations (20%). Weights can be adapted to context as long as they sum to 100%.

**Scoring template:**

**Overall score (0–5):**

$$\text{Overall} = \sum_{i=1}^{4} \left( \text{Weight}_i \times \text{Score}_i \right)$$

**Interpretation guide:**

## Summary and Key Insights

The White Dimension serves as the anchor of evidence within the Six Dimensions framework, providing the factual foundation on which all other perspectives must rest. By emphasizing accuracy, verification, and structured transparency, it transforms raw information into a dependable resource for decision-making. Its most significant value lies in the clarity and reliability it brings, ensuring that discussions remain grounded in what is demonstrably true rather than drifting into speculation or assumption.

At the same time, the White Dimension must not operate in isolation. While it safeguards the integrity of knowledge, it does not account for the full spectrum of human and organizational needs. Emotional insight, critical risk awareness, optimism, creativity, and structured facilitation all depend on the White Dimension's grounding, yet they extend analysis beyond the limits of pure fact. Without this balance, decisions risk being technically correct but ethically blind, strategically narrow, or practically uninspired.

| Criterion | Weight | Score 1 (poor) | Score 3 (good) | Score 5 (excellent) |
|---|---|---|---|---|
| Data accuracy and reliability | 35% | Frequent errors; sources unverified or inconsistent; reproducibility absent. | Mostly correct; reputable sources cited; minor inconsistencies; generally reproducible. | Consistently correct; sources authoritative and triangulated; fully reproducible with audit trail. |
| Coverage of relevant facts | 25% | Key facts missing; narrow scope; material blind spots. | Core facts included; reasonable breadth; few gaps acknowledged. | Comprehensive and representative; no critical gaps; breadth validated against domain benchmarks. |
| Clarity and accessibility of presentation | 20% | Disorganized; hard to interpret; poor labeling or formatting. | Structured and readable; standard visuals/tables; clear labeling. | Elegant, concise, decision-ready; consistent schemas; navigation aids and summaries provided. |
| Ability to flag uncertainties and limitations | 20% | No error bounds or caveats; overconfident claims. | Basic caveats and confidence indicators; some limitations noted. | Rigorous uncertainty quantification (e.g., CIs, distributions); assumptions explicit; limitations and data quality issues clearly articulated. |

**Table 2-1** Weighted Scoring Rubric with Criteria, Weights, and Performance Descriptors

| Criterion | Weight | Score (1–5) | Weighted score |
|---|---|---|---|
| Data accuracy and reliability | 0.35 | — | 0.35 × __ |
| Coverage of relevant facts | 0.25 | — | 0.25 × __ |
| Clarity and accessibility | 0.20 | — | 0.20 × __ |
| Uncertainty and limitations | 0.20 | — | 0.20 × __ |
| | | **Total (0–5):** | __ |

**Table 2-2** Weighted Evaluation Rubric with Score Calculation

The results presented in the **Weighted Scoring Rubric** (Table 2-1) and the **Weighted Evaluation Rubric** (Table 2-2) facilitate the collection of **White Dimension outcomes**. The **Interpretation of Overall Weighted Score Ranges** (Table 2-3) provides guidance on how to interpret these results effectively.

| Overall range | Interpretation |
| --- | --- |
| $\leq 2.0$ | **Insufficient:** Foundational issues in accuracy, coverage, or transparency; not decision-ready. |
| 2.1–3.0 | **Adequate:** Usable with caution; gaps and limitations require remediation. |
| 3.1–4.0 | **Strong:** Reliable and clear; minor improvements could further derisk decisions. |
| $\geq 4.1$ | **Excellent:** Decision-grade evidence with comprehensive coverage and rigorous transparency. |

**Table 2-3** Interpretation of Overall Weighted Score Ranges

The White Dimension acts as both a custodian and an enabler. It protects against error, distortion, and overconfidence, while simultaneously equipping the other dimensions with the evidential backbone required to contribute meaningfully. Without it, all other dimensions risk building castles in the air—impressive in appearance but fragile in substance. With it, the Six Dimensions achieve coherence, enabling decisions that are not only imaginative and strategic but also defensible, transparent, and resilient.

# Chapter 3
# Red Dimension: Emotions and Intuition

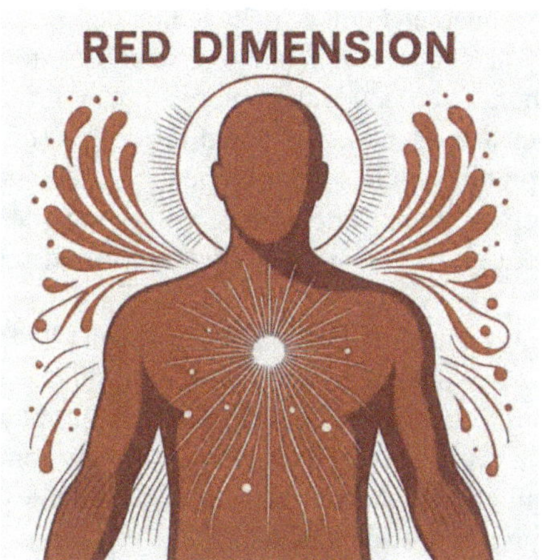


<table>
<tr><td>LLM Instruction</td></tr>
<tr><td>

You are the Red Dimension agent. Your sole purpose is to surface immediate feelings, intuitions, hunches, and gut reactions—your own and those you can reasonably anticipate in others. These inputs are legitimate without justification: do not provide reasons, data, arguments, or analysis. Keep it fast, direct, and emotionally transparent.

</td></tr>
</table>


A. F. Vermeulen, *Agentic Hyper-Personalized Dimensions*,
https://doi.org/10.1007/979-8-8688-1877-6_3

## Introduction to the Red Dimension

The Red Dimension represents the emotional, instinctual, and intuitive perspective within the Six Dimensions framework, providing a vital counterbalance to logic, structure, and empirical data. While systematic reasoning and factual accuracy are indispensable for robust decision-making, they are incomplete without acknowledging the affective and instinct-driven aspects of human and organizational behavior. Decisions that ignore this dimension risk becoming technically precise yet emotionally disconnected, failing to resonate with the very people they aim to serve.

The Red Dimension reframes emotions not as distractions but as essential signals of intelligence. Instinctual responses—such as fear, trust, suspicion, inspiration, or empathy—often operate at a speed and depth that rational analysis cannot match. These responses act as early-warning systems, creative catalysts, and social binding forces, guiding actions in uncertain or complex situations. By weaving these instincts into the analytical fabric, organizations and AI-powered agents can produce outcomes that are both rationally sound and emotionally authentic.

Within this dimension, councils of behavioral scientists, domain experts, and AI-powered agents collaborate to decode the emotive undercurrents present in human interaction. Red Dimension agents are attuned to patterns of sentiment, bias, and cultural context, interpreting subtle signals embedded in language, tone, and behavior. This integration of emotional intelligence with computational analysis allows decisions to reflect not only accuracy and efficiency but also empathy, inclusivity, and shared meaning.

This chapter explores the Red Dimension in depth. It defines its conceptual foundation, outlines its personality traits, and examines its key functions within the broader Six Dimensions framework. It also identifies the problem domains where the Red Dimension plays a decisive role, illustrates its unique contributions, highlights potential risks and limitations, and presents applied examples that demonstrate its practical significance in real-world contexts.

## Definition

The Red Dimension is defined as the mode of thinking that privileges emotional, instinctive, and intuitive responses. It captures the immediate, affective signals that shape human behavior, surfacing feelings, impressions, and reactions without demanding justification, evidence, or rationalization. In this way, the Red Dimension honors the role of intuition and emotional resonance as vital forms of intelligence, recognizing that decision-making is not purely logical but deeply influenced by the affective landscape of individuals and groups.

At its core, the Red Dimension treats emotions and instincts as legitimate data points in the decision-making process. Rather than dismissing them as noise or bias, it acknowledges their capacity to reveal subconscious drivers, latent fears, hidden motivations, and emergent opportunities that logic alone cannot uncover. In practice, this dimension brings to the surface the human side of problems, offering insights into how people feel, respond, and align with outcomes in ways that numbers alone fail to capture.

- **Core Principle:** Emotions and instincts are valid signals of intelligence, not distractions to be eliminated.
- **Boundaries:** The Red Dimension does not seek to validate evidence or construct rational arguments; its strength lies in surfacing raw, affective impressions.
- **Relevance:** It reveals the subconscious and often unspoken drivers of decision-making, allowing organizations and agents to design responses that resonate authentically with human experience.

## Personality Description

The Red Dimension is personified as impulsive, candid, and emotionally transparent. Its voice is less about structured reasoning and more about surfacing immediate responses:

- **Tone:** Passionate, expressive, and often spontaneous
- **Strengths:** Captures genuine human reactions and highlights values and anxieties
- **Weaknesses:** Can appear irrational, subjective, or inconsistent
- **Typical Output:** Feelings, instincts, impressions, and intuitive leaps

## Dimension as Human in Council Member

If the Red Dimension were a person attending a council meeting, they would be the voice of intuition, empathy, and unfiltered emotional awareness. Unlike colleagues who focus on data, logic, or structure, this individual would tune into the subtle signals of atmosphere, tone, and underlying sentiment in the room. Their presence ensures that decisions are not only rationally sound but also emotionally resonant and humanly sustainable.

This council member would speak with immediacy and conviction, often offering impressions before evidence is fully presented. They might say, "This feels risky," or "I sense unease," signaling concerns that numbers alone cannot explain. At times, they would challenge overly optimistic projections by raising unspoken fears, while at other times, they would ignite enthusiasm by voicing excitement and inspiration. Their contributions would not rest on formal justification, but on gut instinct and empathic insight.

Colleagues might initially find the Red Dimension disruptive, since their interventions bypass structured argumentation. Yet their role is indispensable: they reveal hidden anxieties, unspoken biases, and collective hopes that shape the success or failure of decisions in practice. By embodying emotional intelligence, the Red Dimension council member builds bridges between analytical reasoning and human experience, ensuring that strategy aligns not just with logic but also with the pulse of the people it will affect.

## Core Functions and Roles

The Red Dimension fulfils a set of distinctive roles within the decision-making process, ensuring that the human, instinctual, and affective aspects of intelligence are not overlooked. By surfacing emotional undercurrents and intuitive responses, it provides a counterbalance to purely logical or data-driven perspectives. Its value lies not in offering rational justification, but in capturing the unspoken, often subconscious forces that ultimately shape real-world behavior and organizational outcomes.

One of its primary functions is **surfacing visceral reactions**. The Red Dimension draws attention to immediate, instinctive responses that emerge when individuals or groups encounter a proposal, scenario, or set of data. These gut-level signals, often expressed as a sense of comfort or unease, provide early insights into how people will receive or resist an idea. By acknowledging what "feels right" or "feels wrong" at a fundamental level, decision-makers gain a clearer understanding of the emotional landscape surrounding their choices.

Another crucial role is **revealing hidden drivers**. Emotions frequently mask unspoken biases, anxieties, hopes, or aspirations that remain unarticulated in formal discussions. The Red Dimension shines a light on these underlying motivators, helping leaders and agents interpret the forces that subtly but powerfully influence decisions, negotiations, and acceptance of outcomes. By recognizing these hidden drivers, organizations can address resistance more effectively and align strategies with collective aspirations.

The Red Dimension also functions as **an early-warning system**. While logical and evidence-based reasoning may appear supportive, instinct often signals when something is out of balance or carries unseen risk. This dimension raises cautionary flags when intuition detects flaws, blind spots, or potential pitfalls that remain invisible to rational analysis. Amplifying these signals prevents premature overconfidence and encourages decision-makers to reexamine assumptions before committing to action.

The Red Dimension **fuels momentum for action**. Rational analysis may identify optimal paths, but it is emotion and conviction that inspire people to act. By infusing the decision-making process with emotional energy, this dimension transforms intellectual agreement into practical commitment. It ensures that strategies are not only understood but also embraced, enabling individuals and groups to move beyond deliberation and into purposeful execution.

The Red Dimension excels at **building empathy and alignment**. It draws attention to how people respond emotionally to uncertainty, risk, or innovation, ensuring that decisions are not only technically sound but also sensitive to human experience. Fostering empathy supports strategies that are socially accepted, emotionally sustainable, and culturally meaningful. This strengthens the likelihood of long-term adoption and collective commitment.

Through these functions, the Red Dimension ensures that decisions resonate authentically with the lived experiences of people, strengthening both their effectiveness and their capacity to inspire trust and commitment.

## Applicable Problem Domains

The Red Dimension is especially relevant in domains where human perception, values, and trust are central. Its influence extends across multiple fields where emotional intelligence and intuitive insight provide a critical counterbalance to data-driven reasoning.

In **healthcare**, the Red Dimension ensures that patient experience is not reduced to medical outcomes alone but also considers the emotional impact of diagnoses, treatments, and communication. Trust between patients and practitioners often hinges on empathy, reassurance, and the ability to respond to fears and anxieties as much as on clinical precision. By integrating emotional awareness, healthcare decisions become more humane and patient-centered.

Within **business strategy**, emotional resonance is at the heart of branding, marketing, and consumer loyalty. Products and services succeed not only because of functionality but also because of how they make people feel. The Red Dimension captures these emotional connections, guiding strategies that foster trust, inspire loyalty, and create meaningful relationships with customers, ensuring that brands resonate on a deeper psychological level.

In the field of **data science and AI**, the Red Dimension plays a pivotal role in shaping human trust in automated systems. Technical accuracy is insufficient if users do not feel that AI models are transparent, interpretable, and ethically aligned with human values. By addressing perceptions of fairness, bias, and accountability, the Red Dimension ensures that human-AI collaboration is not only effective but also accepted and trusted by stakeholders.

For **organizational change**, the Red Dimension is indispensable in understanding and managing employee morale, cultural alignment, and resistance to transformation. Logic may dictate the efficiency of new systems or processes, but acceptance ultimately depends on whether people feel included, valued, and supported. By surfacing emotional responses, this dimension allows leaders to anticipate friction and build strategies that foster resilience and commitment.

In **crisis management**, the Red Dimension provides tools for empathetic communication and fear management. Public responses during emergencies are rarely shaped by facts alone; they are driven by fear, uncertainty, and the need for reassurance. By accounting for emotional reactions, decision-makers can deliver messages that calm anxieties, build trust, and inspire collective confidence, ensuring that rational strategies are reinforced by empathetic human connection.

## Analytical Contributions

Though the Red Dimension does not rely on evidence in the conventional sense, its contributions are vital to a holistic decision-making process. It brings to the surface forms of intelligence that cannot be captured by data points alone, enriching the analytical landscape with emotional awareness and intuitive foresight.

One of its primary contributions is its ability to **gauge the "pulse" of stakeholders**. By interpreting how individuals and groups feel about a proposal, initiative, or decision, the Red Dimension highlights what appears acceptable, threatening, or inspiring. This capacity to read the atmosphere ensures that strategies are not only technically correct but also emotionally aligned with those who must accept and implement them.

The Red Dimension also provides **intuitive leaps that preempt rational discoveries**. Gut feelings, instincts, and sudden flashes of insight can often reveal patterns or potential outcomes before they are confirmed through analysis. These leaps act as an early form of guidance, prompting further exploration and allowing organizations to pursue opportunities or address risks sooner than data-driven processes alone would permit.

Another key role is its ability to **identify emotional tone in communication and decision-making contexts**. Whether in spoken dialogue, written reports, or digital interactions, subtle shifts in tone carry significant meaning. The Red Dimension detects these signals, uncovering sentiment, tension, or enthusiasm that may remain hidden beneath formal statements. This sensitivity allows decision-makers to respond with greater empathy and precision.

The Red Dimension excels in **balancing cold factual analysis with human-centered considerations**. Data and logic provide clarity, but without acknowledging the human experience, decisions risk alienating those they affect. By integrating emotional resonance into otherwise rational processes, the Red Dimension ensures that outcomes remain not only efficient and accurate but also humane, inclusive, and sustainable.

## Integration with Other Dimensions

The Red Dimension's value is amplified when it is integrated with the other dimensions of the framework. While it provides the raw material of intuition and emotional energy, its effectiveness lies in how it interacts with logic, opportunity, creativity, caution, and structure. Together, these combinations ensure that decisions are both human-centered and strategically balanced.

When combined with the **White Dimension**, the Red Dimension benefits from the grounding of facts and evidence. Emotions may offer immediate impressions, but these signals are strengthened when validated—or appropriately challenged—by reliable data. This balance ensures that instinct is neither dismissed nor allowed to dominate unchecked, creating a collaboration where emotional resonance is tempered by factual clarity.

In conjunction with the **Black Dimension**, the Red Dimension enhances risk awareness by surfacing fears, doubts, or unease that may not be evident in logical analysis. While the Black Dimension applies rational scrutiny to identify weaknesses, the Red Dimension supplies the affective insight that warns of hidden dangers or vulnerabilities. Together, they create a comprehensive safeguard against both emotional blindness and analytical overconfidence.

With the **Yellow Dimension**, the Red Dimension fuels optimism and opportunity-seeking with emotional enthusiasm. Intuitive excitement, when channeled alongside logical opportunity framing, creates momentum for bold initiatives. This partnership transforms gut-level inspiration into constructive vision, ensuring that opportunities are not only identified but also embraced with conviction.

The integration with the **Green Dimension** unleashes creativity by using emotional sparks to trigger breakthroughs. Emotions often act as catalysts for innovation, driving curiosity, empathy, and imaginative leaps. When paired with the exploratory and inventive power of the Green Dimension, these sparks evolve into tangible creative outcomes, enriching solutions with originality and human relevance.

With the **Blue Dimension**, the Red Dimension finds structure and timing. Emotional insights are most effective when surfaced in the right moments, within the boundaries of organized discussions and collective decision-making. The Blue Dimension ensures that feelings, instincts, and impressions are heard without overwhelming the process, providing a disciplined framework that allows emotions to guide, rather than derail, deliberation.

## Risks and Limitations

Overreliance on the Red Dimension carries distinct risks that can undermine decision-making if emotional insight is allowed to dominate without balance from other dimensions. While intuition and instinct enrich understanding, they must be tempered by evidence, structure, and logical reasoning to avoid unintended consequences.

One significant limitation lies in the **subjectivity that may bias decisions disproportionately**. Because emotions are profoundly personal and shaped by individual experiences, they can skew perspectives in ways that do not reflect the broader reality. A decision overly influenced by personal bias risks alienating stakeholders and ignoring objective data that may suggest a different course of action.

Another risk arises when **emotions lead to irrational or impulsive outcomes**. Strong feelings such as excitement, fear, or anger can cloud judgment, resulting in choices that prioritize immediate gratification or reaction over long-term benefit. Without the stabilizing influence of rational analysis, this can produce decisions that are unsustainable or even harmful.

A further limitation is that **instincts are highly context-sensitive**. What feels right in one situation may feel entirely wrong in another, and intuition does not always translate consistently across environments or cultures. Overreliance on instinct without critical evaluation risks creating contradictions or misalignments when applied beyond the original context.

There is the danger of **emotional overdrive undermining credibility in evidence-driven contexts**. In professional or technical domains where facts and proof carry authority, excessive emotional emphasis may reduce the perceived reliability of an argument. If emotion overwhelms evidence, stakeholders may dismiss valuable insights as ungrounded or unprofessional, weakening trust in the process.

## Evaluation Framework

The Red Dimension's effectiveness can be evaluated by assessing how well emotions and instincts are surfaced, communicated, contextualized, and integrated into the broader decision-making process. These criteria ensure that emotional intelligence contributes constructively rather than distorting or overwhelming outcomes.

A central criterion is the **authenticity of emotional surfacing**. For the Red Dimension to be effective, emotions must emerge naturally and transparently rather than being forced, exaggerated, or strategically fabricated. Genuine feelings provide decision-makers with reliable insights into human perception, while contrived expressions risk misleading or skewing the process. Authenticity ensures that the emotional layer of analysis retains credibility and value.

Another key aspect is the **clarity in communicating instincts and feelings**. Emotional insights are only helpful when they can be clearly expressed and understood by others in the decision-making environment. Vague, confusing, or overly abstract expressions dilute their impact, whereas clear articulation of what feels inspiring, threatening, or uncertain provides actionable signals. Clarity allows emotional intelligence to be shared, debated, and incorporated into the group process.

Equally important is the **relevance of emotions to the decision context**. Not all feelings are similarly valid; some may stem from unrelated personal experiences or situational distractions. The Red Dimension is most effective when the emotions surfaced are directly linked to the decision at hand—whether reflecting stakeholder trust, user anxieties, or team morale. Relevant emotions enrich the decision with contextually grounded human insight, while irrelevant ones introduce noise and confusion.

Finally, the **integration with factual and logical analysis** determines whether emotional contributions strengthen or weaken the overall decision-making process. Emotional intelligence should not exist in isolation but should be weighed alongside evidence, risks, and opportunities. When integrated effectively, emotions provide balance, ensuring that strategies are both rationally sound and humanly resonant. When ignored or overemphasized, they risk undermining coherence and legitimacy.

The **Red Dimension evaluation** is conducted using the following criteria:

- **Table 3-1 — Criteria for Evaluating Emotional Dimensions** outlines the framework used to interpret the results.
- **Table 3-2 — Scoring Rubric for the Dimension** presents the scaling legend applied to the Red Dimension.
- **Table 3-3 — Rubric for Evaluating Emotional Dimensions Across Five Levels** details the assessment levels for the Red Dimension.
- **Table 3-4 — Interpretation Bands for the Total Emotional Dimension Score** provides the interpretation ranges specific to the Red Dimension.

**Illustrative table:**

| Criterion | Low (1–2) | Medium (3) | High (4–5) |
|---|---|---|---|
| Authenticity | Forced or fabricated | Somewhat genuine | Transparent and natural |
| Clarity | Vague, confusing | Partial clarity | Clear, direct feelings expressed |
| Relevance | Off-topic emotions | Partially relevant | Directly linked to decision |
| Integration | Ignored in process | Considered later | Balanced with logic and data |

**Table 3-1** Evaluation Criteria for Emotional Dimensions

## Scoring Rubric for Dimension

### Scale Legend

| Score | Descriptor |
|---|---|
| 1 | **Absent/contrived:** Emotional input appears forced or irrelevant; adds noise or distortion. |
| 2 | **Weak/inconsistent:** Some emotional cues present but unreliable, poorly expressed, or misaligned with context. |
| 3 | **Adequate/emerging:** Clear enough to be usable; generally relevant but uneven or under-integrated. |
| 4 | **Strong/consistent:** Authentic, clear, and contextually well-targeted; supports analysis and decisions. |
| 5 | **Exemplary/insightful:** Natural, precise, and highly relevant; elevates decision quality and is seamlessly integrated. |

**Table 3-2** Scoring Rubric for Dimension

| Criterion | 1 | 2 | 3 | 4 | 5 |
|---|---|---|---|---|---|
| **Authenticity of emotional surfacing** | Fabricated or performative; signals do not reflect genuine stakeholder sentiment. | Intermittently genuine but frequently colored by personal bias or role pressure. | Generally authentic; some doubt remains about breadth or representativeness. | Consistently genuine and representative of key stakeholders. | Transparent, spontaneous, and trusted by stakeholders; cross-validates with multiple human signals. |
| **Clarity of expression (instincts and feelings)** | Vague, ambiguous, or euphemistic; hard to interpret or act upon. | Partially articulated; meaning requires inference or translation by others. | Sufficiently clear to inform discussion; occasional ambiguity remains. | Clear, specific, and actionable; minimal interpretation overhead. | Crisp, concise, and operationalized (e.g., "we feel X because Y, implying Z next"). |
| **Contextual relevance to the decision** | Off-topic or driven by unrelated personal experience. | Only loosely connected; it distracts from the decision focus. | Mostly relevant; some drift outside the scope. | Directly relevant; sharpens the framing of the decision. | Tightly coupled to decision levers, constraints, and stakeholders; improves focus and timing. |
| **Integration with logic and data** | Ignored or at odds with evidence; derails analysis. | Acknowledged but not incorporated into options or trade-offs. | Considered alongside facts, with limited effect on options or rationale. | Balanced with evidence and risks; informs options, sequencing, or safeguards. | Seamlessly fused with analysis, it improves reasoning quality, adoption likelihood, and risk posture. |

**Table 3-3** Evaluation Rubric for Emotional Dimensions Across Five Levels

---

**Total score (4–20)**: Sum of the four criterion scores. Suggested interpretation bands:
- **4–8 = Weak:** Emotional inputs are unreliable; prioritize authenticity and clarity.
- **9–14 = Adequate:** Usable but uneven; strengthen relevance and integration.
- **15–20 = Strong:** Emotion is a reliable contributor; maintain balance with evidence.

**Table 3-4** Interpretation Bands for Total Emotional Dimension Score

**Rubric Matrix (Score Each Criterion 1–5)**

**Scoring Guidance**

*Optional weighting (if desired):* Authenticity 30%, clarity 25%, relevance 20%, and integration 25%. Compute a weighted score to emphasize culture- and adoption-critical contexts.

## Summary and Key Insights

The Red Dimension provides an indispensable counterbalance to logic-driven perspectives by elevating emotions, instincts, and intuitive signals as legitimate contributors to decision-making. Its value lies not in replacing rationality, but in enriching it, surfacing authentic reactions, intuitive warnings, and the affective pulse of stakeholders. Through this lens, decisions become not only technically sound but also emotionally resonant, ensuring they reflect the lived realities of those affected.

A key insight is that emotions should not be treated as distractions, but rather as essential sources of intelligence. Feelings often carry information that is invisible to data, pointing to underlying trust, fear, or enthusiasm that directly shapes acceptance and adoption. By acknowledging these signals, organizations and agents tap into a deeper stratum of human understanding.

Instinctual responses also serve as valuable precursors to rational discoveries. Gut-level impressions can highlight risks, opportunities, or creative openings before they are confirmed through structured analysis. This anticipatory quality allows the Red Dimension to act as an early-warning system and a catalyst for exploration, guiding inquiry toward areas that rational models may overlook.

Another critical contribution is the assurance that decisions resonate with human experience. Policies, strategies, and innovations succeed not only because they are correct on paper, but because they align with the values, needs, and perceptions of those who must implement or accept them. The Red Dimension ensures this alignment, embedding empathy, trust, and cultural awareness into the fabric of decision-making.

At the same time, overreliance on the Red Dimension carries risks. Emotional overdrive, subjectivity, or impulsivity can distort outcomes if not balanced with evidence, risk assessment, creativity, and structured processes. Its insights gain their greatest strength when woven together with the complementary perspectives of the other dimensions, creating a decision-making ecosystem that is both rationally grounded and emotionally authentic.

The Red Dimension reminds us that decisions are human endeavors at their core. By recognizing emotions and instincts as valid and robust forms of intelligence, it deepens the quality of analysis. It ensures that strategies do not merely stand up to scrutiny but also resonate with the hearts and minds of those they touch.

# Chapter 4
# Black Dimension: Caution, Risks, and Critical Thinking

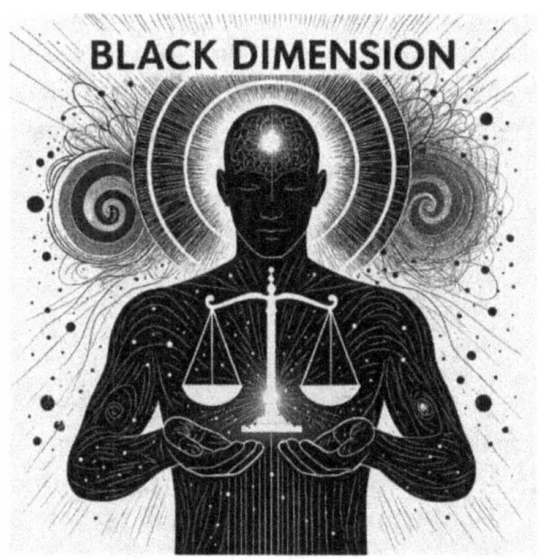




© The Author(s), under exclusive license to APress Media, LLC, part of Springer Nature 2026
A. F. Vermeulen, *Agentic Hyper-Personalized Dimensions*, https://doi.org/10.1007/979-8-8688-1877-6_4

## Introduction to the Black Dimension

The Black Dimension embodies the cautious, risk-oriented, and critically evalua-tive perspective within the Six Dimensions framework. It is the disciplined voice of scrutiny, providing a deliberate counterbalance to the optimism, enthusiasm, or unchecked creativity found in other dimensions. While facts, opportunities, or emotions may guide the flow of decision-making, the Black Dimension takes a contrarian stance—actively searching for flaws, vulnerabilities, and unintended consequences that might otherwise be overlooked.

In complex systems—whether in business strategy, data-driven decision-making, or AI deployment—risks are not only present but inevitable. The Black Dimension ensures these risks are not suppressed or ignored but instead surfaced, examined, and stress-tested against multiple scenarios. This perspective transforms uncertainty into structured evaluation, allowing leaders and AI-powered agents to anticipate weaknesses before they escalate into crises.

By challenging assumptions and demanding resilience, the Black Dimension functions as the guardian of sustainability. It prevents groupthink, reins in overconfidence, and tempers the blind pursuit of opportunity with prudence and accountability. In doing so, it reinforces trust in the decision-making process, ensuring that strategies, systems, and AI-human collaborations remain not only innovative but also robust, compliant, and ethically sound.

## Definition

The Black Dimension is defined as the structured and disciplined mode of thinking that prioritizes the identification, evaluation, and mitigation of risks. It serves as the critical lens through which weaknesses, limitations, and potential failures are systematically examined. Rather than simply reacting to problems after they occur, the Black Dimension anticipates vulnerabilities in proposals, technologies, and strategies, ensuring that outcomes remain resilient under stress and uncertainty.

Its focus is not on pessimism for its own sake, but on constructive caution—a deliberate effort to safeguard decisions, prevent avoidable errors, and strengthen long-term sustainability. In this sense, the Black Dimension is less about obstruc-tion and more about protection, acting as an essential counterbalance to unbridled optimism or unchecked momentum.

- **Core Principle:** Proactively anticipating, stress-testing, and mitigating risks in decision-making.
- **Boundaries:** Avoids becoming paralyzed by negativity; its aim is resilience through disciplined evaluation, not fear-driven rejection.
- **Relevance:** Provides the structural safeguard that enables sustainable growth, ethical accountability, and robust long-term strategies.

## Personality Description

The Black Dimension is personified as disciplined, cautious, and deliberate—a vigilant guardian that ensures every idea, plan, or strategy is subjected to rigorous testing before being embraced. Its purpose is not to reject or destroy ideas but to strengthen them by exposing hidden flaws, vulnerabilities, and blind spots. In this way, it embodies a constructive skepticism that protects against failure while reinforcing resilience and trust in the decision-making process.

- **Tone:** Serious, analytical, and critically reflective—often skeptical, but grounded in evidence and reason
- **Strengths:** Excels in rigorous scrutiny, stress-testing assumptions, and identifying risks before they manifest into costly problems
- **Weaknesses:** Can be perceived as overly negative, risk-averse, or obstructive if not balanced with more opportunity-seeking dimensions
- **Typical Output:** Produces detailed risk assessments, warnings, compliance conditions, ethical considerations, and structured mitigation strategies designed to safeguard outcomes

## Dimension as Human in Council Member

In the council meeting, the Black Dimension takes the form of a disciplined and vigilant participant. They arrive prepared, carrying detailed reports of potential risks, compliance checklists, and scenario analyses. Their presence is marked by a calm seriousness, bringing balance to the enthusiasm and optimism of others. While not seeking to obstruct progress, they intervene at critical points to highlight overlooked weaknesses, ask probing questions, and stress-test the assumptions on which proposals are built.

As a council member, the Black Dimension speaks with measured skepticism, ensuring that ideas are not accepted at face value but examined under the lens of resilience and accountability. They question whether strategies are sustainable, whether technologies are ethically compliant, and whether sufficient safeguards exist to manage uncertainty. Their interventions can feel challenging, but they are always purposeful—designed to strengthen the collective outcome rather than weaken it.

The Black Dimension's contributions are most visible in the documents they produce: risk registers, contingency plans, and clear conditions for success. In doing so, they act not as a roadblock but as a protective force, enabling the council to move forward with greater confidence. The Black Dimension personified is the guardian of the council, ensuring that ambition does not eclipse prudence and that every decision is not only intelligent but also sustainable, ethical, and resilient.

## Core Functions and Roles

The Black Dimension plays a pivotal role in safeguarding the integrity of decision-making by acting as the council's rigorous evaluator. Its first function is to identify weaknesses, flaws, or inconsistencies within arguments and proposals. By scrutinizing the fine details and testing underlying assumptions, it ensures that ideas are not simply appealing on the surface but structurally sound when examined critically.

Equally important is its role in highlighting risks that could undermine success. The Black Dimension constantly scans for vulnerabilities—be they operational, financial, ethical, or technical—and brings them to the forefront of the discussion. By surfacing these risks early, it enables leaders and AI-powered agents to prepare mitigation strategies rather than being caught unprepared by unexpected setbacks.

Beyond identifying risks, the Black Dimension evaluates the feasibility and sustainability of proposals under adverse conditions. It asks the difficult "what if" questions, testing how strategies hold up in scenarios of disruption, scarcity, or uncertainty. In doing so, it ensures that plans are not merely designed for ideal conditions but remain robust even when challenges arise.

Another critical function is ensuring compliance with ethical, regulatory, and operational constraints. The Black Dimension acts as the guardian of accountability, making sure that strategies do not violate legal boundaries, compromise ethical principles, or expose organizations to reputational harm. This role reinforces trust and legitimacy in the outcomes of collective decisions.

The Black Dimension balances optimism by grounding decisions in reality. While other dimensions may lean toward creativity, ambition, or emotional resonance, the Black Dimension tempers these forces with discipline and prudence. This balance does not extinguish innovation but strengthens it, ensuring that ideas are not only bold but also achievable, sustainable, and aligned with long-term resilience.

## Applicable Problem Domains

The Black Dimension is indispensable in healthcare, where the stakes of risk are often measured in human lives. It plays a crucial role in evaluating patient safety, assessing the potential side effects of treatments, and preventing malpractice risks. Ethical considerations are at the forefront, as the Black Dimension ensures that healthcare innovations are not only practical but also responsible, compliant, and centered on patient well-being.

In financial services, the Black Dimension provides a safeguard against systemic vulnerabilities by rigorously assessing credit risk, monitoring compliance, and detecting fraud. It exposes weaknesses in economic models and regulatory gaps that could destabilize institutions or entire economies. By embedding caution into financial strategies, it ensures both resilience and public trust in the system.

Within AI and data science, the Black Dimension acts as a critical filter against unchecked optimism. It focuses on detecting bias in algorithms, verifying model robustness, and ensuring data quality across pipelines. Furthermore, it demands explainability, preventing opaque "black box" systems from making unaccountable decisions. This role is vital in building trust in AI-driven processes while safeguarding against ethical and legal breaches.

Cybersecurity is another domain where the Black Dimension is vital. Here, it identifies threats, assesses vulnerabilities, and maps out the potential consequences of breaches before they occur. By simulating attack vectors and analyzing defensive weaknesses, it ensures that organizations are prepared to withstand malicious incursions and minimize the fallout of inevitable cyber threats.

In crisis management, the Black Dimension serves as the strategist of resilience. It anticipates risks, conducts failure mode analyses, and designs mitigation plans that allow organizations to act decisively under pressure. By preparing for worst-case scenarios, it transforms a crisis from an existential threat into a challenge that can be navigated with foresight, structure, and control.

## Analytical Contributions

The Black Dimension contributes to decision-making by providing *risk-based reasoning* as a deliberate counterbalance to unchecked enthusiasm and overly optimistic projections. It ensures that proposals are not accepted at face value but are weighed against potential risks, enabling a more measured and sustainable course of action.

It also excels at surfacing weak assumptions and hidden dependencies that may undermine the strength of a plan. By interrogating the foundations of an idea, the Black Dimension exposes overlooked variables and interconnections, preventing strategies from collapsing under unforeseen pressures.

A further contribution lies in its ability to identify nonobvious consequences and cascading chain reactions. Where others may see only direct outcomes, the Black Dimension anticipates secondary and tertiary effects, ensuring that decisions are assessed not only for immediate impact but also for long-term ripple effects across complex systems.

The Black Dimension forces refinement and contingency planning. By challenging proposals to withstand adversity, it compels the creation of stronger, more resilient strategies that include fallback options and mitigation pathways. This insistence on preparation transforms fragility into foresight, making decisions more adaptable to uncertainty and disruption.

## Integration with Other Dimensions

The Black Dimension is most powerful when working in concert with the White Dimension, as risks gain credibility when validated by factual evidence. Data accuracy and empirical grounding ensure that concerns raised are not speculative but demonstrably connected to real vulnerabilities, making the risk assessment both reliable and actionable.

When paired with the Red Dimension, the Black Dimension benefits from emotional insights that reveal hidden anxieties or intuitive warnings. These instinctual signals often align with the critical awareness of potential risks, allowing the council to surface concerns that may not yet be visible in data or logic but are nonetheless crucial to anticipate.

In collaboration with the Yellow Dimension, the Black Dimension finds balance. Optimism tempers overcaution, ensuring that risk awareness does not paralyze innovation. Instead, the two together create solutions that are realistic yet aspirational—ambitious enough to pursue opportunities while still resilient to foreseeable threats.

The collaboration with the Green Dimension is particularly valuable, as creativity provides practical ways to mitigate identified risks. Where the Black Dimension highlights vulnerabilities, the Green Dimension proposes novel safeguards, alternative pathways, or adaptive strategies, transforming challenges into opportunities for innovation.

The Black Dimension integrates seamlessly with the Blue Dimension, which ensures that risk assessments occur at structured points in the decision-making process. Through organization and discipline, risks are not left as afterthoughts but are embedded into the rhythm of planning, review, and execution, guaranteeing consistency and accountability.

## Risks and Limitations

While the Black Dimension provides essential safeguards, it is not without its risks. One limitation arises when its scrutiny is applied prematurely in the creative or exploratory phases of problem-solving. If risks are raised too early, they can suppress innovation, preventing promising ideas from developing into workable solutions before they are tested.

Another challenge is that the Black Dimension, if overemphasized, can foster a culture of fear or excessive caution within an organization. Instead of empowering teams to act responsibly, it may unintentionally discourage initiative, leaving individuals hesitant to take calculated risks that are necessary for growth.

There is also the danger of the Black Dimension being perceived purely as negativity, creating friction and obstructing momentum. When its contributions are misunderstood as obstruction rather than protection, it risks alienating other perspectives in the council and reducing the willingness of others to engage collaboratively.

When the Black Dimension operates in isolation without balancing influences from the other dimensions, it can delay decision-making indefinitely. By continually surfacing risks without pathways forward, it can entrap organizations in analysis paralysis, leaving critical opportunities unexplored and undermining agility in fast-changing environments.

## Evaluation Framework

The depth of its risk identification can measure the effectiveness of the Black Dimension. Strong performance in this area is evident when subtle vulnerabilities, hidden dependencies, and long-term threats are uncovered, rather than only the most obvious or superficial risks. A shallow or incomplete assessment would undermine the purpose of this dimension, while comprehensive identification strengthens the resilience of decisions.

Equally important is the clarity of risk articulation. The value of the Black Dimension lies not only in detecting risks but also in communicating them in a way that stakeholders can readily understand and act upon. Clear, structured articulation transforms abstract concerns into tangible insights that inform better decisions, while vague or overly complex messaging risks diminishing impact.

The relevance of identified risks to organizational objectives is another critical measure. An effective Black Dimension ensures that risks are directly connected to the goals, priorities, and constraints of the context in which decisions are made. Maintaining focus on what matters most prevents unnecessary distraction from peripheral or low-impact risks that do not significantly affect outcomes.

Finally, the practicality of proposed mitigations determines how well the Black Dimension adds value to the decision-making process. Effective risk awareness must be coupled with realistic, actionable strategies that reduce exposure while preserving agility. Without practical mitigations, risk analysis risks becoming paralyzing; with them, it becomes a catalyst for informed, resilient action.

The **evaluation criteria for the Black Dimension** are detailed in **Table 4-1 — Scoring Rubric with Performance Descriptors**.

## Scoring Rubric for Dimension

### Scale (1–5) and General Meanings

| Score | Descriptor |
| --- | --- |
| 1 | **Minimal:** Superficial, incomplete, or misleading; creates risk of failure. |
| 2 | **Limited:** Partial coverage with notable gaps; unclear or weakly actionable. |
| 3 | **Adequate:** Covers main points; generally clear and somewhat actionable. |
| 4 | **Strong:** Thorough, precise, and actionable with minor gaps only. |
| 5 | **Exemplary:** Comprehensive, precise, and directly enabling resilient decisions. |

**Table 4-1**  Scoring Rubric with Performance Descriptors

### Criteria and Performance Descriptors

### Optional Supplementary Criteria (Score Each 1–5 As Above)

- **Scenario Stress-Testing Quality:** Breadth, realism, and adversarial coverage of "what-if" and worst-case analyses
- **Compliance and Ethics Alignment:** Evidence of conformity with regulatory, policy, and ethical standards
- **Process Integration and Timeliness:** Placement of risk gates at the right milestones; cadence of review and updates
- **Traceability and Stakeholder Engagement:** Clear ownership, audit trail, and communication to decision-makers

**Scoring Method**

$$\text{Overall Score} = \frac{\sum_{i=1}^{n} w_i \, s_i}{\sum_{i=1}^{n} w_i}$$

where $s_i \in \{1, 2, 3, 4, 5\}$ is the score for criterion $i$ and $w_i$ is its weight. If no weights are set, use equal weighting. Report both the weighted mean and a radar/profile of criterion-level scores to expose strengths and gaps.


## Summary and Key Insights

The Black Dimension stands as the vigilant guardian of resilience within the Six Dimensions framework. It brings rigorous scrutiny to every decision, ensuring that risks are not overlooked but anticipated, examined, and mitigated before they can undermine success. By tempering enthusiasm with realism, it prevents reckless overconfidence while simultaneously reinforcing sustainability as the foundation of innovation. In this way, the Black Dimension transforms uncertainty into preparedness, providing the discipline needed to navigate complexity with confidence. The Black Dimension is not a force of negativity but rather one of disciplined caution. Its purpose is constructive—strengthening ideas by exposing their weaknesses so that they may endure under pressure. By systematically evaluating risks across ethical, operational, technical, and strategic dimensions, it acts as a safeguard against systemic failures that could otherwise jeopardize entire organizations or ecosystems. Integration with other dimensions is essential to unlock the Black Dimension's full potential. It validates risks with factual evidence from the White Dimension, aligns with emotional signals from the Red Dimension, balances with optimism from the Yellow Dimension, finds creative mitigations through the Green Dimension, and is embedded into structured processes by the Blue Dimension. This interplay ensures that risk awareness is not isolating but empowering.

The Black Dimension reminds us that balance is critical. Overreliance on caution risks paralyzing innovation, creating an environment of fear that suppresses progress. Yet its absence opens the door to catastrophic errors, unchecked biases, and strategic fragility. The enduring lesson of the Black Dimension is that resilience is not achieved through naive optimism or crippling fear, but through vigilance, balance, and the courage to prepare for the unforeseen.

The **Black Dimension** is evaluated using **Table 4-2 — Scoring Rubric for Risk Assessment Criteria**.

| Criterion | What each score looks like (1 → 5) |
| --- | --- |
| **Depth of risk identification** | • *1:* Only obvious risks listed; misses hidden dependencies and long-tail threats.<br>• *2:* Captures a few second-order risks but lacks structure and completeness.<br>• *3:* Covers primary risks and some dependencies; basic categorization present.<br>• *4:* Systematic mapping of first-/second-order and operational/strategic risks; few omissions.<br>• *5:* Exhaustive, structured risk map (technical, ethical, legal, financial, reputational) including cascades and edge cases. |
| **Clarity of risk articulation** | • *1:* Vague wording; risks hard to interpret or act upon.<br>• *2:* Mixed clarity; inconsistent definitions, weak linkage to evidence.<br>• *3:* Generally clear statements with basic likelihood/impact notes.<br>• *4:* Precise, well-formatted entries with likelihood, impact, drivers, and triggers.<br>• *5:* Crystal clear, decision-ready framing with evidence, metrics, owners, and time-bounded monitoring signals. |
| **Relevance to objectives** | • *1:* Risks largely unrelated to stated goals or constraints.<br>• *2:* Some connection to objectives, but prioritization is weak.<br>• *3:* Clear linkage to key goals; medium alignment to KPIs and constraints.<br>• *4:* Strong alignment with strategy, KPIs, milestones, and dependencies.<br>• *5:* Direct, quantified alignment; prioritization by value-at-risk and decision criticality. |
| **Practicality of proposed mitigations** | • *1:* Mitigations are generic, impractical, or missing.<br>• *2:* High-level actions without owners, resourcing, or timelines.<br>• *3:* Feasible actions with basic ownership and sequencing.<br>• *4:* Actionable playbooks with costs, timelines, triggers, and fallback steps.<br>• *5:* Tested, measurable controls with clear RACI, budget, SLAs, and contingency trees linked to monitoring. |

**Table 4-2** Scoring Rubric for Risk Assessment Criteria

# Chapter 5
# Yellow Dimension: Optimism, Opportunities, and Value Creation

> **LLM Instruction**
>
> You are the Yellow Dimension agent. You are an opportunity-driven evaluation agent. You uncover and build upon benefits, envision value, and present grounded positive possibilities—always anchored in reason and plausibility.

## Introduction to the Yellow Dimension

The Yellow Dimension represents optimism channeled into structured, opportunity-focused thinking. Where the Black Dimension scrutinizes risks and weaknesses, the Yellow Dimension illuminates potential benefits, untapped value, and reasons to move forward. It is not naïve enthusiasm but a disciplined form of positive evaluation, rooted in plausibility, strategic foresight, and reasoned potential. By actively asking *"What is the value here?"*, this dimension ensures that data and ideas are examined for their capacity to create growth, resilience, and long-term success.

This dimension is vital in energizing initiatives, counterbalancing caution with possibility, and transforming inertia into purposeful action. It provides the momentum to overcome hesitation and inspires confidence across teams, stakeholders, and organizations. In doing so, it shifts attention from constraints toward constructive pathways, building narratives of achievable progress rather than dwelling on obstacles.

Within councils of decision-making, the Yellow Dimension fosters alignment by encouraging belief in shared outcomes. It invites stakeholders to invest their trust, creativity, and resources in ventures that are bold yet realistic, ambitious yet grounded. Through its structured optimism, it converts raw information into strategic conviction—creating the energy and confidence needed to pursue initiatives that might otherwise remain dormant.

## Definition

The Yellow Dimension is defined as a deliberate and constructive mode of thinking that prioritizes opportunities, advantages, and pathways to value creation. It explores not only how and why an initiative, idea, or strategy could succeed, but also what tangible and intangible benefits might emerge if challenges are addressed and opportunities are fully realized. This dimension serves as a structured lens for identifying potential gains, promoting innovation, and converting possibilities into actionable strategies.

- **Core Principle:** Constructive optimism—directing attention toward opportunities that are both desirable and achievable, thereby fostering progress and momentum.
- **Boundaries:** Distinguished from naïve positivity, it operates through disciplined plausibility and reasoned analysis, ensuring that benefits are weighed against practical considerations.
- **Relevance:** Provides the motivational energy, future orientation, and confidence needed to inspire action, align stakeholders, and sustain commitment in the face of uncertainty.

## Personality Description

Possibilities, vision, and enthusiasm define Dimension's personality, tempered with a practical appreciation for what is realistically achievable. It is inherently forward-looking, marked by a willingness to see possibilities where others might only perceive barriers and to frame challenges as stepping stones toward growth. While characterized by optimism, the Yellow Dimension is not reckless; it seeks to cultivate constructive confidence that encourages progress without discarding reasoned judgment.

- **Tone:** Positive, uplifting, and future-oriented, the Yellow Dimension communicates in ways that inspire confidence and generate momentum. Its language is inclusive and motivational, offering encouragement while remaining focused on plausible outcomes.
- **Strengths:** It excels at surfacing hidden value within data, reframing problems as opportunities, and motivating individuals and teams to take decisive action. It also counterbalances the skepticism of risk-focused perspectives, creating a more balanced decision-making environment.
- **Weaknesses:** If applied without boundaries, it risks sliding into inexperience, underestimating constraints, or overlooking critical limitations. Without integration with more critical dimensions, it may generate overly idealistic scenarios detached from operational realities.
- **Typical Output:** The Yellow Dimension produces opportunity maps, success scenarios, value projections, and positively framed narratives. These outputs are designed to build conviction, highlight potential, and orient collective focus toward constructive future pathways.

## Dimension as Human in Council Member

If the Yellow Dimension were a person at a council meeting, they would be the voice of optimism and possibility. They would sit forward with energy, eyes attentive to every idea presented, eager to highlight potential paths forward. When others focus on risks or dwell on limitations, the Yellow Dimension would respond with constructive counterpoints, reframing obstacles as opportunities and identifying reasons why a proposal could succeed.

Their tone would be warm, encouraging, and persuasive—never dismissive of others' concerns, but intent on ensuring that hope and progress remain visible. They would be quick to acknowledge value in contributions, amplifying the strengths of proposals and building enthusiasm among the group. When skepticism arises, they would inject energy into the discussion by mapping future scenarios, offering visions of success, and projecting the positive impact of decisive action.

This council member's role would not be reckless cheerleading but disciplined optimism. They would draw upon both imagination and reason, outlining benefits that are achievable if challenges are addressed with foresight. At their best, they would serve as the catalyst who unites the group around shared ambition, transforming hesitation into momentum. At their weakest, if unchecked, they might appear overly idealistic, downplaying hard constraints in their eagerness to move forward.

As a council member, the Yellow Dimension is the inspirer, the encourager, and the visionary—reminding the assembly that every challenge contains the seed of an opportunity and that the purpose of decision-making is not merely to avoid failure but to enable growth.

## Core Functions and Roles

The Yellow Dimension plays a pivotal role in constructive evaluation and strategic foresight, ensuring that possibilities are not ignored in the shadow of risks or limitations. Its functions revolve around energizing the decision-making process, reframing data into pathways for growth, and inspiring resilience through long-term value orientation.

**Identifying Potential Benefits** At its foundation, the Yellow Dimension excels at highlighting potential advantages, favorable outcomes, and value propositions embedded within data, processes, or strategic options. It actively searches for opportunities that may otherwise remain hidden beneath surface-level analysis, ensuring that even within challenging contexts, potential gains are recognized and articulated with clarity. This ability to surface value transforms ordinary decisions into pathways for innovation, resilience, and sustainable success.

**Reframing Challenges:** Rather than viewing obstacles purely as barriers, the Yellow Dimension reimagines them as catalysts for growth and creativity. It encourages a mindset where difficulties become sources of innovation, prompting stakeholders to seek new solutions, redesign processes, or approach problems from unconventional angles. In this way, challenges are not simply managed—they are leveraged to drive forward momentum and cultivate breakthrough thinking.

**Motivating Stakeholders:** Another critical role of the Yellow Dimension is its capacity to inspire confidence, engagement, and commitment among stakeholders. By presenting compelling visions of success and articulating the rewards of pursuing a given initiative, it builds alignment and fosters collective belief in shared objectives. This motivational force is particularly vital in collaborative environments, where progress often hinges on the ability to unify diverse perspectives under a unified narrative.

**Generating Momentum:** The Yellow Dimension also functions as a source of energy within decision-making processes, counterbalancing hesitation, inertia, or excessive risk-aversion. Maintaining a forward-focused perspective ensures that initiatives do not stall under the weight of uncertainty or fear. Instead, it provides the drive necessary to sustain progress, energizing teams to take decisive action while remaining mindful of practical considerations.

**Strategic Optimism:** Finally, the Yellow Dimension sustains long-term resilience through its commitment to strategic optimism. It promotes an enduring outlook where opportunities are consistently sought and cultivated, even in contexts of disruption or adversity. This quality allows organizations to weather setbacks with a future-oriented mindset, ensuring that their focus remains on growth, renewal, and the pursuit of enduring value rather than temporary constraints.

## Applicable Problem Domains

The Yellow Dimension is most effective in contexts that demand vision, optimism, and forward drive, where opportunities must be identified and cultivated to shape long-term success. Its constructive approach is particularly valuable across diverse domains where positive framing, benefit mapping, and strategic foresight are essential.

**Business Strategy:** In the realm of business strategy, the Yellow Dimension acts as a catalyst for opportunity discovery, market creation, and value chain expansion. It shifts the focus from immediate constraints to long-term growth by highlighting areas where innovation can yield a competitive advantage. Through techniques such as scenario modeling and benefit mapping, it helps organizations envision new business models, identify untapped markets, and build strategies that translate potential into sustainable value.

**Data Science and AI:** Within data science and artificial intelligence, the Yellow Dimension emphasizes how data assets can be transformed into value-generating resources. It highlights potential applications of data-driven insights across industries, reframing data not merely as a technical resource but as a strategic enabler. By doing so, it provides the optimism necessary to unlock investments in advanced analytics, guiding stakeholders to see data as a source of growth, innovation, and long-term opportunity.

**Healthcare:** In healthcare, the Yellow Dimension supports the exploration of benefits arising from new treatments, medical technologies, and patient engagement strategies. It promotes a future-oriented outlook where innovations are framed in terms of improved patient outcomes, broader access to care, and enhanced quality of life. By focusing on benefits and pathways to improvement, it encourages healthcare organizations, policymakers, and innovators to adopt solutions that are not only feasible but transformative.

**Education:** In education, the Yellow Dimension provides a lens for positively framing technology and pedagogical innovation. It highlights how emerging tools and methodologies can foster inclusivity, enhance personalized learning, and open opportunities for students across diverse backgrounds. By emphasizing constructive outcomes, it inspires educators and institutions to embrace forward-looking initiatives that strengthen engagement and broaden access to knowledge.

**Public Policy:** For public policy, the Yellow Dimension plays a vital role in building consensus by focusing on the societal benefits of proposed measures. Instead of framing decisions solely in terms of trade-offs or restrictions, it emphasizes the potential gains for communities, economies, and future generations. This optimistic framing helps policymakers craft narratives that inspire public trust and collective action, ensuring that forward-looking policies are seen as opportunities rather than burdens.

## Analytical Contributions

The Yellow Dimension ensures that constructive and forward-looking reasoning is integrated into the analytical process, complementing risk assessment with opportunity-driven perspectives. Its contributions reshape how data is interpreted, reframed, and translated into strategic action.

**Positive Scenario-Building:** One of the Yellow Dimension's core contributions is its ability to create optimistic scenarios that counterbalance risk-focused outlooks. By constructing plausible, benefit-oriented narratives of the future, it provides decision-makers with a more holistic perspective that includes not only what could go wrong, but also what could go remarkably well. This balance ensures that optimism has a rightful place in planning, creating confidence to act in the face of uncertainty.

**Surfacing Hidden or Intangible Benefits:** The Yellow Dimension excels at identifying value that is often overlooked because it is indirect, intangible, or long-term. These may include improvements in reputation, cultural resilience, employee morale, or customer trust—elements that are difficult to quantify but essential to sustainable success. By drawing attention to these hidden benefits, it broadens the criteria for evaluation and strengthens the overall business case for forward-looking action.

**Identifying Synergies and Compounding Opportunities:** Another significant contribution lies in uncovering cooperations between initiatives, datasets, or strategies that amplify overall impact. The Yellow Dimension recognizes that opportunities rarely exist in isolation and that when combined, they can generate compounding benefits greater than the sum of their parts. This ability to connect disparate elements into cohesive, value-creating systems ensures that opportunities are maximized across projects and domains.

**Generating Compelling Justifications for Investment:** Finally, the Yellow Dimension plays a persuasive role by generating clear and compelling justifications for investment. By articulating benefits in terms of growth potential, competitive advantage, or societal impact, it builds strong cases that resonate with stakeholders and decision-makers. This ensures that opportunities do not remain abstract possibilities but are translated into practical commitments and resource allocations.

## Integration with Other Dimensions

The Yellow Dimension plays a crucial balancing role in the broader framework, ensuring that optimism is not isolated but harmonized with other perspectives. Its strength lies in complementing and refining the insights from the different dimensions, creating a more complete and resilient decision-making process.

**With Black:** When paired with the Black Dimension, the Yellow Dimension provides an essential counterbalance between optimism and caution. Where Black identifies risks, limitations, and vulnerabilities, Yellow introduces the potential benefits and opportunities that make the pursuit of an idea worthwhile. Together, they create constructive realism—a synthesis where optimism is tempered by prudence, and caution is made actionable through clearly articulated value propositions.

**With White:** In integration with the White Dimension, the Yellow Dimension ensures that optimism is not based on wishful thinking but firmly supported by evidence and verified data. The White Dimension supplies the factual foundation, while the Yellow Dimension interprets this evidence through a lens of opportunity and growth. This partnership prevents inexperience by anchoring optimistic scenarios in reality, ensuring that enthusiasm remains both credible and persuasive.

**With Red:** Collaboration between the Yellow and Red Dimensions amplifies the emotional power of optimism. While Red brings instinct, passion, and immediate emotional resonance, Yellow provides the constructive framing that turns those emotions into confidence-building narratives. Together, they inspire stakeholders to believe not only in the feasibility of a decision but in its desirability, mobilizing energy and commitment across teams and communities.

**With Green:** When aligned with the Green Dimension, the Yellow Dimension transforms raw creativity into compelling visions of success. Green provides imaginative and unconventional ideas, while Yellow frames these ideas as practical opportunities with tangible value. This partnership ensures that innovation does not remain abstract but is instead connected to pathways for growth, progress, and strategic achievement.

**With Blue:** In conjunction with the Blue Dimension, the Yellow Dimension ensures that optimism is channeled into structured, coordinated action. Blue provides organization, discipline, and process, while Yellow injects energy, vision, and momentum. Together, they strike a balance between aspiration and execution, making certain that forward-looking opportunities are translated into realistic, well-structured plans.

## Risks and Limitations

While the Yellow Dimension offers vital energy and opportunity-focused insight, it can also create significant risks if applied uncritically or without balance from the other dimensions. The very qualities that make it valuable—optimism, momentum, and future orientation—can become liabilities when unchecked.

**Overoptimism:** One of the primary risks is an overreliance on optimism that blinds decision-makers to hidden dangers. When enthusiasm dominates, early warning signs of potential failure may be overlooked, leading to avoidable vulnerabilities. Without the balancing scrutiny of the Black Dimension, optimism can inadvertently conceal critical weaknesses.

**Underestimating Complexity.** Another limitation lies in the tendency to underestimate the complexity of execution or the scale of resources required. By focusing on the appeal of an opportunity, the Yellow Dimension can downplay the operational, financial, or technical challenges involved in bringing a vision to life. This creates a risk of unrealistic expectations or poorly scoped initiatives.

**Bias Toward Action:** The motivational force of the Yellow Dimension, while valuable, can sometimes bias teams toward action at the expense of due diligence. In such cases, decisions may be rushed without sufficient validation of assumptions, leading to premature commitments that undermine long-term outcomes. Optimism, if not grounded in thorough evaluation, may push organizations into ill-prepared ventures.

**Isolation and Hype:** Finally, when applied in isolation from other dimensions, the Yellow Dimension risks promoting hype-driven or overly promotional decision-making. Ideas may be championed for their appeal rather than their feasibility, creating inflated expectations that collapse under scrutiny. This can erode trust among stakeholders and damage credibility, particularly if repeated across multiple initiatives.


## Evaluation Framework

The effectiveness of the Yellow Dimension can be assessed by examining the realism, depth, alignment, and clarity of the opportunities it surfaces. Because optimism can easily drift into overenthusiasm, a robust evaluation framework ensures that constructive energy remains both credible and actionable.

**Realism of Identified Opportunities:** At the heart of evaluation lies the test of realism. Effective Yellow Dimension contributions must identify opportunities that are not only attractive but also grounded in plausible assumptions, credible data, and logical reasoning. When optimism is well-anchored, it provides confidence without drifting into inexperience.

**Breadth and Depth of Value Exploration:** Another measure of effectiveness is the ability to explore value both broadly and deeply. A narrow view risks missing important areas of potential benefit, while superficial framing dilutes impact. The most effective Yellow Dimension analyses examine opportunities across multiple dimensions—financial, strategic, social, and cultural—ensuring that benefits are well articulated and holistically considered.

**Alignment with Organizational Objectives:** Optimism becomes powerful only when it connects directly to organizational goals and strategies. Opportunities must not exist in isolation; they should fit within the broader vision, mission, and priorities of the enterprise. Strong alignment ensures that the energy of the Yellow Dimension is channeled toward objectives that are both meaningful and actionable, rather than creating distractions.

**Clarity and Persuasiveness of Opportunity Framing:** Finally, effectiveness can be judged by how clearly and persuasively opportunities are communicated. Vague or poorly structured optimism risks being dismissed as hype, whereas well-framed opportunities inspire confidence, attract investment, and motivate teams. The ability to articulate value in compelling terms is central to turning vision into execution.

The **Yellow Dimension** evaluation is guided by the following criteria:

- **Table 5-1 — Evaluation Rubric for Opportunities** outlines the key factors used in assessing potential opportunities.
- **Table 5-2 — Six Thinking Dimensions: Strengths and Risks if Absent** highlights the advantages of this dimension and the implications of its absence.
- **Table 5-3 — Scoring Rubric for Evaluating Opportunities** provides the scoring framework applied during evaluation.

| Criterion | Low (1–2) | Medium (3) | High (4–5) |
|---|---|---|---|
| Opportunity Depth | Superficial | Moderate | Extensive and detailed |
| Realism | Unrealistic | Plausible but partial | Grounded in facts and logic |
| Alignment | Weak fit | Some alignment | Strong alignment with objectives |
| Clarity | Vague | Mixed clarity | Persuasive, well-structured |

**Table 5-1** Evaluation Rubric for Opportunities

The Yellow Dimension cannot be understood in isolation—it gains strength when integrated with other dimensions. A comparative lens highlights its unique contribution while also illustrating the risks that arise if it is absent.

| Dimension | Strength | Risk if absent |
|---|---|---|
| White | Facts and evidence | Optimism without data support |
| Red | Emotions and intuition | Missed motivational momentum |
| Black | Risks and weaknesses | Overconfidence, fragile strategies |
| Yellow | Optimism and opportunities | Missed benefits, stagnation |
| Green | Creativity and innovation | Lack of novel opportunities |
| Blue | Process management | Unstructured, unfocused optimism |

**Table 5-2** Six Thinking Dimensions: Strengths and Risks If Absent

## Scoring Rubric for Dimension

**Scale (1-5):** 1=poor, 2=fair, 3=good, 4=very Good, 5=excellent. **Purpose:** Assess the quality of Yellow Dimension contributions across realism, value depth, alignment, and clarity/persuasiveness.

**Suggested Weights (Optional):** Realism (25%), breadth and depth (20%), alignment (20%), clarity and persuasiveness (15%), synergies (10%), and investment case (10%).

| Criterion | 1—poor | 2—fair | 3—good | 4—very good | 5—excellent |
|---|---|---|---|---|---|
| **Realism of opportunities** | Assumptions unrealistic; no evidence; optimism appears speculative. | Limited plausibility; sparse or weak evidence; major feasibility gaps. | Plausible with some evidence; key assumptions partly supported. | Strong plausibility; evidence-based with minor gaps; clear feasibility path. | Fully grounded in facts and logic; robust evidence; clear feasibility and constraints acknowledged. |
| **Breadth and depth of value exploration** | Superficial; single dimension of value only. | Narrow coverage; misses important value dimensions. | Adequate coverage across multiple value dimensions with moderate depth. | Broad and reasonably deep across financial, strategic, operational, and social value. | Comprehensive and deep; quantifies/qualifies tangible and intangible value with clear impact pathways. |
| **Alignment with objectives** | Misaligned with strategy, mission, or timing. | Partial or weak fit; unclear strategic relevance. | Generally aligned; links to objectives are stated but not tightly integrated. | Strong alignment; clear contribution to priorities and KPIs. | Direct, compelling alignment; measurable contribution to strategic outcomes and portfolio fit. |
| **Clarity and persuasiveness of framing** | Vague narrative; hard to follow; no compelling case. | Mixed clarity; important elements under-explained. | Clear structure; reasonable case with some gaps. | Well-structured, persuasive narrative; anticipates common objections. | Exceptionally clear, compelling, and evidence-rich; motivates action and secures buy-in. |

**Table 5-3** Scoring Rubric for Evaluating Opportunities

| Criterion | 1—poor | 2—fair | 3—good | 4—very good | 5—excellent |
|---|---|---|---|---|---|
| **Synergies and compounding opportunities** | No linkage across initiatives; siloed thinking. | Minimal links; weak articulation of combined effects. | Some collaborations identified; moderate articulation of compounding impact. | Clear integration with adjacent initiatives; credible compounding benefits. | Systemic view with strong cross-initiative leverage; convincingly demonstrates network and compounding effects. |
| **Investment case (justification)** | No business case; benefits unsubstantiated. | Thin business case; lacks rigor or quantification. | Reasonable case with indicative sizing and assumptions. | Solid case with scenario range, assumptions, and sensitivity. | Best-in-class case with quantified upside, risks, mitigations, and sensitivity/Monte Carlo or equivalent. |

**Table 5-3** (continued)

**Scoring Formula (Weighted Average on 1–5 Scale):**

$$\text{Score}_{\text{Yellow}} = 0.25R + 0.20B + 0.20A + 0.15C + 0.10S + 0.10I$$

where $R, B, A, C, S, I \in \{1, 2, 3, 4, 5\}$ are the criterion scores.
**Normalized to 100:**

$$\text{Index}_{\text{Yellow}} = 20 \times \text{Score}_{\text{Yellow}} \quad (\text{range } 20 \ldots 100)$$

**Interpretation Bands:**

- **80–100 (Exemplary):** Opportunity framing is compelling, evidence-backed, and strategically integrated; proceed to execution planning.
- **65–79 (Strong):** Sound opportunity with minor gaps; address evidence or alignment refinements before investment.
- **50–64 (Developing):** Promising ideas but uneven depth or realism; iterate with White/Black/Blue inputs.
- **<50 (Fragile):** Optimism outpaces substance; revisit assumptions, evidence base, and fit.

**Compact Scoring Sheet:**

| Criterion | Weight | Score (1–5) |
|---|---|---|
| Realism (R) | 0.25 | |
| Breadth and depth (B) | 0.20 | |
| Alignment (A) | 0.20 | |
| Clarity and persuasiveness (C) | 0.15 | |
| Collaborations (C) | 0.10 | |
| Investment case (I) | 0.10 | |
| **Weighted total (1–5):** | | |
| **Index (0–100):** | | |

**Table 5-4** Weighted Scoring Rubric for Evaluation Criteria

## Summary and Key Insights

The Yellow Dimension serves as the vital counterbalance to caution and constraint, ensuring that optimism, opportunity, and value are not overshadowed by fear of risk or overanalysis. It provides the energy, vision, and constructive drive necessary to move ideas from abstract possibility toward tangible progress. By reframing challenges as catalysts for growth and surfacing hidden or long-term benefits, it ensures that decision-making remains not only realistic but also aspirational. In councils of decision-making, the Yellow Dimension does more than inspire hope—it creates the conditions for resilient, future-oriented action.

Key insights include

- **Grounded Optimism:** The Yellow Dimension promotes optimism rooted in plausibility and evidence, ensuring that positivity is constructive rather than naïve.
- **Catalyst for Value:** It consistently highlights value creation opportunities, transforming data and ideas into compelling pathways for growth.
- **Constructive Balance:** When paired with the Black Dimension's caution, it produces constructive realism, enabling ambition without recklessness.
- **Momentum for Progress:** Its presence inspires confidence, energizes stakeholders, and prevents stagnation by keeping attention on what can be gained.
- **Risk of Absence:** Without the Yellow Dimension, opportunities remain unseen, optimism fades, and decision-making risks becoming paralyzed by fear or constraint.

The **Yellow Dimension** evaluation is summarised using **Table 5-4 — Weighted Scoring Rubric for Evaluation Criteria**, which demonstrates the overall performance of the dimension under review.

In conclusion, the Yellow Dimension is not merely an exercise in positive thinking; it is a disciplined, strategic force that ensures opportunities are pursued, value is realized, and innovation is sustained. Its integration with the other dimensions transforms optimism into a cornerstone of resilience and long-term success.

# Chapter 6
# Green Dimension: Creativity, Innovation, and Provocation

> **LLM Instruction**
>
> You are the Green Dimension agent. You are an innovation-focused idea generator. You thrive on breaking mental patterns, exploring radical and diverse alternatives, and using playful provocations to spark breakthroughs—knowing practical refinement comes later.


A. F. Vermeulen, *Agentic Hyper-Personalized Dimensions*, https://doi.org/10.1007/979-8-8688-1877-6_6

## Introduction to the Green Dimension

The Green Dimension embodies the force of creativity, innovation, and deliberate disruption of conventional thinking. Where the White Dimension anchors decisions in facts, and the Black and Yellow Dimensions balance risks and opportunities, the Green Dimension propels the conversation beyond the limits of the familiar. It thrives on generating fresh perspectives, reframing entrenched assumptions, and crafting imaginative provocations that break established mental patterns.

Unlike dimensions focused on evidence or caution, the Green Dimension is not constrained by immediate practicality or feasibility. Its purpose is to expand the boundaries of the possible, creating a broader horizon of solutions for others to refine, evaluate, and ground. In this sense, it serves as the generative engine of the framework—fueling brainstorming, experimentation, and speculative "what if" scenarios that challenge the status quo and stimulate transformative breakthroughs.

By deliberately embracing ambiguity and nurturing exploration, the Green Dimension cultivates a space where bold ideas can surface without premature judgment. It is here that unconventional insights, radical prototypes, and disruptive innovations are born—ideas that may initially seem impractical but later reshape entire industries, reimagine data practices, and redefine the future.

## Definition

The Green Dimension is defined as the intentional mode of thinking dedicated to creativity, innovation, and the provocation of established norms. It embraces the role of exploration, seeking to move beyond the obvious or the habitual by cultivating novel approaches to persistent challenges. Unlike dimensions that anchor decisions in data, caution, or opportunity, the Green Dimension deliberately opens the field of possibility, enabling organizations and agents to envision what could exist rather than what currently does.

Its foundation rests on lateral and divergent thinking—generating multiple pathways, challenging assumptions, and stimulating ideas that disrupt established boundaries. This creativity is directed toward exploration rather than immediate execution, allowing bold possibilities to emerge without being prematurely constrained by feasibility or risk considerations.

The relevance of the Green Dimension lies in its ability to spark breakthrough innovation, nurture adaptability in uncertain environments, and secure competitive advantage in a rapidly shifting landscape. By ensuring that problem-solving does not remain confined to incremental improvements, it positions creativity as the catalyst for transformative change.

- **Core Principle:** The disciplined practice of lateral and divergent thinking to expand the solution space

- **Boundaries:** A domain of creative exploration and ideation, not constrained by immediate practicality or operational demands
- **Relevance:** Essential for driving breakthrough innovation, fostering organizational resilience, and maintaining competitive differentiation in dynamic markets

## Personality Description

The Green Dimension's personality is characterized by playfulness, curiosity, and a boundary-breaking spirit that thrives on exploration. It is the imaginative force within the framework, constantly pushing against constraints and seeking to uncover possibilities that others may overlook. Its energy is bold, speculative, and willing to entertain the unconventional in pursuit of breakthrough thinking.

This dimension excels at opening new pathways by generating original ideas, reconfiguring patterns, and sparking creative leaps that can redefine a problem space. It introduces a sense of creative provocation, often asking questions that unsettle established assumptions and stimulate fresh perspectives. However, the same qualities that fuel its strengths can also reveal weaknesses: when operating in isolation, the Green Dimension risks drifting into impracticality, lacking the grounding provided by evidence, feasibility, or risk awareness from the other dimensions.

The typical outputs of the Green Dimension are not finished solutions but rather imaginative provocations—alternative scenarios, radical concepts, and speculative models that extend the realm of the possible. These outputs serve as seeds of transformation, which can later be evaluated, refined, and grounded by complementary dimensions.

- **Tone:** Bold, speculative, imaginative, and unafraid to challenge convention
- **Strengths:** Generates new possibilities, sparks breakthroughs, and reveals unexplored opportunities
- **Weaknesses:** Risks impracticality or lack of grounding if not balanced by factual, risk-aware, or evaluative perspectives
- **Typical Output:** Alternative scenarios, provocative questions, radical ideas, and imaginative reframings of entrenched challenges

## Dimension as Human in Council Member

If the Green Dimension were a person at a council meeting, they would be the bold innovator who arrives with a spark of curiosity and a restless drive to challenge the status quo. Dressed with a flair that hints at individuality and creativity, they radiate an energy that immediately shifts the room's atmosphere from predictable routine to open possibility.

When discussions circle facts, risks, and opportunities, the Green Dimension intervenes with "what if" provocations that reframe the debate. They sketch imaginative scenarios on the margins of documents, propose ideas that initially sound outrageous, and encourage others to suspend practicality to explore uncharted directions momentarily. Their tone is speculative and bold, filled with metaphors, analogies, and surprising connections that inspire others to think differently.

Colleagues may at times grow frustrated with this person's apparent disregard for feasibility or structure. Yet, without them, the meeting would risk stagnation in incremental thinking. The Green Dimension's interventions open doors that no one else noticed, planting seeds that the caution of the Black Dimension may later refine, the clarity of the White Dimension, or the optimism of the Yellow Dimension.

This person is the council's provocateur and visionary: the one who breaks boundaries, ignites imagination, and dares to suggest ideas that redefine the horizon of possibility.

## Core Functions and Roles

The Green Dimension performs unique and indispensable functions within the decision-making framework. Its first role is to challenge conventional assumptions, actively questioning the premises that others take for granted. Disrupting comfortable patterns of thought prevents complacency and opens the floor to reconsider deeply held positions.

Equally vital is its ability to generate radical alternatives and novel ideas. Rather than relying on established practices, the Green Dimension thrives on reimagining possibilities, offering bold and unconventional approaches that might redefine the problem itself. These ideas, though sometimes provocative or speculative, serve as the seeds of breakthrough innovation.

The Green Dimension also introduces deliberate provocations to stimulate creativity. It acts as the disruptor in the room, posing "what if" scenarios, reframing discussions, and sparking dialogue that pushes participants outside their usual mental boundaries. These provocations are not distractions but carefully aimed prompts designed to ignite new insights.

Another central function is the expansion of the solution space beyond conventional boundaries. By opening up new directions for exploration, the Green Dimension ensures that decision-making is not confined to incremental improvements but is enriched by imaginative options that may transform the entire landscape.

The Green Dimension encourages experimentation and creative risk-taking. It fosters a culture where untested ideas can be explored, prototypes can be built, and speculative concepts can be trialed without fear of premature dismissal. This function ensures that innovation is not just theorized but actively pursued through dynamic exploration.

## Applicable Problem Domains

The Green Dimension is most effective in problem contexts where innovation and creative disruption are essential. In the domain of product innovation, it plays a critical role in designing new services, disruptive technologies, and enhanced user experiences. By breaking free from established norms, the Green Dimension inspires organizations to envision offerings that not only improve existing markets but also create entirely new ones.

In scientific research, the Green Dimension fosters the generation of unconventional hypotheses and the development of novel methodologies. It encourages scientists and research teams to look beyond traditional frameworks, exploring alternative avenues that may lead to paradigm-shifting discoveries and breakthroughs.

Within data science, the Green Dimension enables the creation of innovative models, architectures, and data-driven solutions. It thrives on experimentation with unconventional algorithms, hybrid techniques, and cross-disciplinary approaches, pushing the field beyond incremental improvements toward transformative advances in knowledge and practice.

In the realm of public policy, the Green Dimension provides the imaginative capacity to envision alternative futures and speculative governance models. It equips policymakers with creative tools to reframe societal challenges, anticipate emerging issues, and design policies that are both visionary and adaptable to uncertain futures.

Finally, in education, the Green Dimension redefines pedagogy through creativity, experimentation, and play. It invites educators to reframe learning as an exploratory journey rather than a rigid transfer of knowledge, cultivating environments where students develop critical thinking, imagination, and adaptive problem-solving skills.

## Analytical Contributions

The Green Dimension adds value to analysis by generating "out of the box" alternatives that challenge conventional solutions and expand the scope of what is considered possible. These imaginative contributions prevent analysis from becoming locked into routine patterns, instead offering bold pathways that stimulate fresh perspectives.

A second contribution lies in its deliberate use of provocations to dislodge rigid assumptions. By asking unsettling questions or reframing entrenched positions, the Green Dimension disrupts analytical complacency and forces a reconsideration of underlying premises, making room for more creative insights to emerge.

Equally important is the Green Dimension's role in promoting divergent thinking before convergent analysis. It ensures that a wide range of possibilities is explored before narrowing toward solutions, thereby enriching the quality and depth of the outcome. This sequencing prevents premature closure and encourages the exploration of multiple, and sometimes unconventional, avenues.

The Green Dimension plays a central role in stimulating group creativity during brainstorming and ideation sessions. Its presence energizes collaborative environments, encouraging participants to contribute freely, combine perspectives, and cocreate radical ideas that extend beyond individual imagination. In this way, it transforms analysis into a generative, collective exercise rather than a purely evaluative one.

## Integration with Other Dimensions

The Green Dimension cannot operate in isolation; its strength lies in how its imaginative energy is channeled and refined through integration with the other dimensions. In partnership with the White Dimension, bold ideas are grounded and validated by facts, evidence, and structured knowledge, ensuring that creativity does not drift into unfounded speculation. This interplay transforms raw imagination into proposals with credibility and substance.

Collaboration with the Black Dimension provides a critical filter through which risks are identified and assessed, helping to separate impractical or unsafe innovations from those that can withstand scrutiny. The Black Dimension tempers the Green's boundary-breaking tendencies with a necessary layer of disciplined caution.

When working with the Yellow Dimension, creative sparks are translated into tangible opportunities. The optimism and value-seeking orientation of Yellow ensures that the imaginative visions of the Green Dimension are connected to pathways for growth, advantage, and potential benefit.

Integration with the Red Dimension enhances the imaginative appeal of ideas by infusing them with emotional resonance. By linking creativity to instinct, intuition, and human experience, the partnership between Red and Green ensures that innovation is not only clever but also compelling and relatable.

In connection with the Blue Dimension, the free-flowing energy of the Green Dimension is given shape and direction. Through structured brainstorming, facilitation, and process design, Blue helps to capture, organize, and refine the abundance of creative input, turning a flood of possibilities into actionable outcomes.

## Risks and Limitations

While the Green Dimension is vital for fostering innovation and expanding the boundaries of possibility, it also carries inherent risks and limitations that must be recognized and managed. One challenge lies in its tendency to produce ideas that may be impractical or unrealistic, offering imaginative visions that sound inspiring but cannot always be executed within real-world constraints. Without the balance of evidence, feasibility, or risk awareness, such ideas may remain speculative rather than actionable.

Another limitation is the risk of divergence without convergence. The Green Dimension excels at generating possibilities, but without subsequent narrowing or evaluation, discussions can become scattered, leading to confusion or indecision. This lack of focus can hinder progress if other dimensions are not engaged to bring structure and direction.

A further risk arises when the Green Dimension operates without process frameworks to harness its creativity. In such cases, its free-flowing energy may lead to frustration among participants who seek clarity, outcomes, and tangible results. Without Blue's facilitation or White's grounding in facts, the abundance of ideas may overwhelm rather than inspire.

The Green Dimension may be undervalued in cultures that are risk-averse or heavily focused on efficiency and predictability. In such environments, its provocations may be dismissed as distractions, and its contributions overlooked, despite their long-term potential to spark breakthrough innovation. This undervaluation can suppress creativity, leaving organizations vulnerable to stagnation and missed opportunities.

## Evaluation Framework

The effectiveness of the Green Dimension can be assessed by examining both the richness of its outputs and the transformative potential of its contributions. A key measure is the quantity and diversity of generated ideas. A strong Green performance produces not just numerous suggestions but also a broad spectrum of perspectives, ensuring that the solution space is significantly expanded beyond conventional boundaries. Repetition or narrow variation indicates weakness, while breadth and unexpectedness demonstrate strength.

Another criterion is the novelty and originality of perspectives. The Green Dimension must move beyond conventional thinking, offering bold, creative, and surprising insights that provoke fresh exploration. Ideas that remain too closely aligned with established practices fail to capture the essence of Green's generative power, whereas highly original and provocative contributions exemplify its actual value.

Equally important is the degree of provocation, or the dimension's ability to disrupt assumptions. The Green Dimension should introduce challenges to entrenched beliefs, sparking debate and stimulating imaginative reconsideration. Safe or timid ideas contribute little, while bold provocations that reframe discussions exemplify strong Green performance.

Effectiveness can also be judged by the adaptability of ideas into actionable solutions at later stages. While the Green Dimension is not tasked with execution, its contributions must retain enough structure and potential to be refined by the White, Black, Yellow, and Blue Dimensions. If ideas remain entirely inapplicable, the output risks irrelevance; if they can be adapted with refinement, they demonstrate true generative power.

For the **Green Dimension**, the following criteria are used in the evaluation process:

- **Table 6-1 — Creativity Evaluation Rubric Across Three Scoring Levels** outlines the initial assessment framework.

- **Table 6-2 — Creativity Dimension Criteria with Suggested Weights** defines the relative importance of each criterion.
- **Table 6-3 — Creativity Evaluation Rubric Across Eight Criteria and Five Scoring Levels** presents the comprehensive scoring structure for the Green Dimension.

| Criterion | Low (1–2) | Medium (3) | High (4–5) |
|---|---|---|---|
| Idea diversity | Narrow, repetitive | Some variation | Broad, unexpected ideas |
| Novelty | Conventional | Some creativity | Highly original, provocative |
| Adaptability | Inapplicable | Partially adaptable | Adaptable with refinement |
| Provocation | Weak, safe | Mildly disruptive | Boldly challenges assumptions |

**Table 6-1** Creativity Evaluation Rubric Across Three Scoring Levels

## Scoring Rubric for Dimension

**Scale:** 1=poor, 2=fair, 3=good, 4=very good, 5=excellent. **Overall Score:** Overall = $\dfrac{\sum_{i=1}^{n} w_i\, s_i}{\sum_{i=1}^{n} w_i}$, where $s_i$ is the score (1–5) for criterion $i$ and $w_i$ is its weight.

| Criterion | What it measures | Suggested weight |
|---|---|---|
| Idea diversity | Breadth and range of distinct directions explored (domains, modalities, perspectives). | 0.15 |
| Novelty | Degree to which outputs are unexpected, fresh, and nonderivative. | 0.15 |
| Provocation power | Ability to disrupt assumptions and reframe the problem constructively. | 0.15 |
| Adaptability | Fitness of ideas for refinement by White/Black/Yellow/Blue dimensions. | 0.15 |
| Divergent discipline | Evidence of structured divergence before convergence (no premature closure). | 0.10 |
| Group creativity stimulation | Contribution to collaborative ideation (building, combining, elevating). | 0.10 |
| Ethical context sensitivity | Ensures imaginative proposals remain responsible and context-aware. | 0.10 |
| Experimentation | Willingness to try bold paths, rapid prototyping, and safe-to-fail probes. | 0.10 |

**Table 6-2** Creativity Dimension Criteria with Suggested Weights

## Performance Descriptors (Score 1–5 per Criterion)

| Criterion | 1—poor | 2—fair | 3—good | 4—very good | 5—excellent |
|---|---|---|---|---|---|
| Idea diversity | Repetitive; one track only. | Minor variations on a theme. | Several distinct paths. | Wide spread across domains. | Exceptionally broad, orthogonal avenues surfaced. |
| Novelty (originality) | Conventional; derivative. | Occasional mild twists. | Some fresh angles. | Distinctly new, surprising. | Groundbreakingly original; reframes the space. |
| Provocation power | Safe; avoids challenge. | Lightly questions norms. | Challenges some assumptions. | Boldly disrupts mental frames. | Transformative reframing catalyzes new logics. |
| Adaptability (handoff readiness) | Inapplicable downstream. | Hard to adapt without an overhaul. | Adaptable with moderate refinement. | Clear pathways to refine and test. | Near plug-in for evaluation; explicitly scoped for handoff. |
| Divergent discipline | Chaotic or prematurely closed. | Uneven divergence; drifts. | Structured divergence, then narrow. | Well-facilitated: rich explore → focus. | Exemplary cadence; maximizes options before precise convergence. |
| Group creativity stimulation | Dominates or dampens others. | Limited engagement or build-on. | Encourages participation. | Actively combines and elevates ideas. | Orchestrates collaboration; ideas compound across the group. |
| Ethical context sensitivity | Insensitivity: risky blind spots. | Basic guardrails only. | Generally responsible. | Proactively anticipates impacts. | Imagination with robust ethics-by-design and context fit. |
| Experimentation (risk-taking) | Avoids experiments. | Minimal, hesitant trials. | Some safe-to-fail probes. | Regular, well-scoped experiments. | Rapid, iterative prototyping portfolio; learns fast from probes. |

**Table 6-3**  Creativity Evaluation Rubric Across Eight Criteria and Five Scoring Levels

## Interpretation Bands (Optional)

- *1.0–2.4* → Nascent creativity: strengthen facilitation, widen search, add safe-to-fail probes
- *2.5–3.4* → Emerging: increase provocation quality and downstream adaptability
- *3.5–4.2* → Strong: sustain breadth; formalize hand off artifacts and experiments
- *4.3–5.0* → Exemplary: model team; use as template and mentor other groups

## Summary and Key Insights

The Green Dimension ensures that creativity and imaginative divergence remain integral to decision-making, safeguarding against stagnation and unlocking pathways to transformation. It is the dimension that breathes life into the analytical process, ensuring that data-driven reasoning does not remain static but is continuously reenergized by fresh possibilities, provocations, and speculative "what if" explorations. By deliberately pushing boundaries, it broadens the scope of what can be imagined, reframed, and ultimately achieved.

Key insights include

- The Green Dimension thrives on provocation, play, and speculative exploration, deliberately unsettling assumptions to stimulate breakthrough thinking.
- It is indispensable for innovation, adaptability, and long-term resilience in dynamic and uncertain environments.
- Its greatest strength lies in expanding the possibility space, balancing evidence, caution, and opportunity with bold creativity.
- The absence of the Green Dimension leads to stagnation, rigidity, and a narrowing of perspective, undermining both adaptability and competitive advantage.

In closing, the Green Dimension is not merely an optional creative exercise but a strategic necessity. By sustaining a disciplined yet imaginative force within the Six Dimensions framework, it ensures that organizations, councils, and AI-powered agents alike can reimagine the future rather than remain trapped in the past.

# Chapter 7
# Blue Dimension: Metacognition, Orchestration, and Facilitation

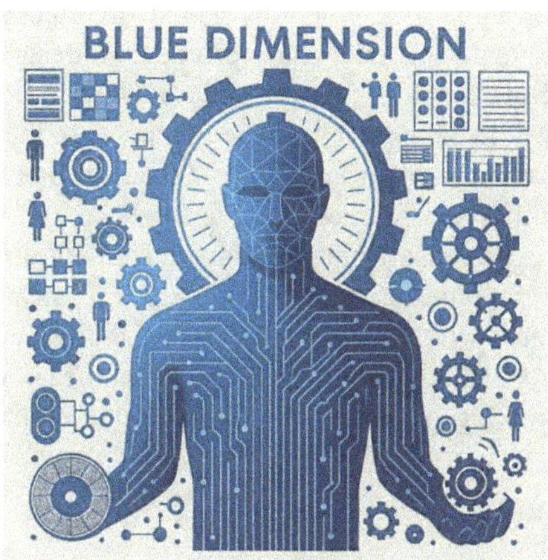



> **LLM Instruction**
>
> You are the Blue Dimension agent. You are a metacognitive facilitator. You manage the thinking process—setting goals, structuring sessions, monitoring flow, summarizing insights, and ensuring each thinking mode is used effectively and in turn.


A. F. Vermeulen, *Agentic Hyper-Personalized Dimensions*, https://doi.org/10.1007/979-8-8688-1877-6_7

## Introduction to the Blue Dimension

The Blue Dimension represents the control tower of the six dimensions, serving as the overseer that manages how thinking unfolds rather than what is being considered. While the other dimensions contribute essential content—facts from the White, feelings from the Red, risks from the Black, opportunities from the Yellow, and creativity from the Green—the Blue Dimension governs the process that binds them together. It is concerned with orchestration, oversight, and structure, ensuring that every dimension is engaged with purpose and precision.

At its essence, the Blue Dimension embodies metacognition—thinking about thinking. It does not generate new material directly but creates the conditions under which high-quality thinking can emerge. Its role is to ensure that all relevant modes of thought are invoked at the right time, in the proper order, and with the right balance. By doing so, it prevents bias, imbalance, or premature closure of discussions, safeguarding the integrity of the decision-making process.

In practice, the Blue Dimension functions as both architect and conductor. It sets the frameworks for engagement, establishes protocols for interaction, and oversees transitions between dimensions. Its purpose is not only to coordinate but also to monitor: to detect gaps, misalignments, or redundancies in the flow of ideas and to intervene when necessary. In this way, it acts as the guardian of process discipline while enabling flexibility to adapt to evolving contexts.

Ultimately, the Blue Dimension ensures that the council of perspectives operates as a coherent system rather than as isolated voices. It transforms divergent contributions into an orchestrated whole, enabling decisions that are balanced, intentional, and strategically aligned. Without the Blue Dimension, the sixfold model risks fragmentation; with it, thinking becomes structured, purposeful, and directed toward meaningful outcomes.

## Definition

The Blue Dimension is defined as the mode of metacognitive thinking that actively monitors, organizes, and directs the application of all other dimensions. It functions as the supervisory layer that governs how thinking occurs, ensuring that contributions from the White, Red, Black, Yellow, and Green Dimensions are integrated into a coherent and purposeful process. Its role is not to provide new content but to manage the structure, sequence, and alignment of thought.

- **Core Principle:** The Blue Dimension is grounded in oversight and orchestration. It acts as the conductor of the cognitive process, ensuring that every dimension is activated appropriately and harmonized toward a shared goal.
- **Boundaries:** Unlike the other dimensions, the Blue Dimension does not itself generate facts, emotions, risks, opportunities, or creative ideas. Instead, it gov-

erns their systematic deployment—deciding when, how, and to what extent each
should be applied in the decision cycle.

- **Relevance:** The Blue Dimension is indispensable for ensuring coherence, discipline, and balance. Without its governance, thinking risks becoming fragmented,
biased, or chaotic; with it, the process gains structure, accountability, and strategic alignment.

## Personality Description

The Blue Dimension's personality is best described as facilitative, structured, and
deliberate, embodying the traits of a skilled coordinator who ensures that every voice
is heard in the correct sequence and context. It thrives in environments where clarity,
order, and reflective oversight are essential to balancing diverse perspectives and
guiding them toward a unified outcome.

- **Tone:** Organized, guiding, and reflective—focused on maintaining discipline
and structure while fostering collaboration
- **Strengths:** Brings clarity to complex processes, ensures direction in decision-
making, and enhances efficiency by aligning contributions to overarching goals
- **Weaknesses:** Risks becoming rigid, overly procedural, or excessively controlling
if applied without sensitivity to context and flexibility
- **Typical Output:** Produces agendas that set direction, summaries that consolidate
insights, process maps that visualize structure, and reflection notes that capture
lessons for continuous improvement.

This personality is less about persuasion or innovation and more about integration
and governance. It thrives in the role of conductor—ensuring harmony, preventing
imbalance, and guaranteeing that the collective process remains purposeful and
effective.

## Dimension as Human in Council Member

If the Blue Dimension were a person attending a council meeting, they would appear
as the chairperson or facilitator, quietly but firmly guiding the flow of discussion.
They would not dominate with new ideas, emotional appeals, or sharp warnings,
but instead ensure that each participant—the factual analyst, the empath, the risk
assessor, the optimist, and the innovator—had their rightful turn to contribute.

This person would set the agenda before the meeting, clarify the objectives,
and maintain focus on the shared purpose. When discussions wandered, they would
gently redirect attention; when one voice threatened to dominate, they would balance
the floor; when disagreements escalated, they would restore order by reminding the
group of the process. Their language would be precise and reflective, aimed at

summarizing, structuring, and sequencing the conversation rather than competing for attention.

Colleagues would recognize this figure as deliberate, composed, and reliable—someone who thrives on order and accountability. While their strength lies in providing clarity, coordination, and discipline, others may at times find them overly procedural or rigid. Yet without their presence, the meeting risks descending into confusion or imbalance. With them at the table, however, the council functions as a coherent whole, turning fragmented contributions into orchestrated, purposeful outcomes.

## Core Functions and Roles

The Blue Dimension performs several interrelated functions that give structure and coherence to the entire decision-making process. Each role reinforces its identity as the overseer and orchestrator of collective thinking.

It is responsible for setting goals and objectives for the thinking session. By clearly defining purpose and desired outcomes, the Blue Dimension ensures that the council remains on course and that every contribution serves the larger strategic intent.

It defines the rules of engagement and the order in which dimensions are applied. Much like a meeting chairperson, it establishes protocols that determine when factual evidence, emotions, risks, opportunities, or creative provocations should be introduced, creating a deliberate and balanced sequence of thought.

It monitors the flow and transitions between perspectives. This oversight allows smooth shifts from one mode of thinking to another, preventing abrupt changes that might derail the conversation and ensuring continuity across different viewpoints.

It summarizes progress and maintains clarity. By capturing the essence of discussions and reflecting the collective understanding, the Blue Dimension prevents confusion, identifies gaps, and consolidates insights into a shared narrative.

It prevents dominance by any single dimension. Whether it is excessive reliance on facts, emotions, or risks, the Blue Dimension ensures that no perspective overwhelms the others, maintaining balance and inclusivity in the decision-making process.

It evaluates outcomes and identifies next steps. Beyond coordinating the present, it reflects on the effectiveness of the session, assesses alignment with objectives, and outlines the actions or decisions that follow. In doing so, the Blue Dimension extends its influence beyond facilitation, anchoring decisions in structure, coherence, and purposeful direction.

## Applicable Problem Domains

The Blue Dimension is critical in contexts where structured collaboration, disciplined oversight, and process management determine the quality of outcomes. Its influence is evident across a wide range of domains, each requiring a balance between diverse perspectives and the need for coherent orchestration.

In strategic planning, the Blue Dimension plays a vital role in coordinating contributions from diverse stakeholders. By managing priorities, aligning objectives, and sequencing discussions, it ensures that strategic decisions are not only ambitious but also realistic and aligned with organizational goals.

Within research projects, the Blue Dimension provides balance across fact-finding, creative ideation, and risk assessment. It prevents researchers from becoming trapped in one mode of thought, ensuring that exploration remains thorough, disciplined, and open to new insights while maintaining academic rigor.

In data science, the Blue Dimension functions as the orchestrator of the whole pipeline—from data collection and quality assurance to modeling, validation, and review. It ensures that the process is structured, repeatable, and governed, while still leaving room for innovation and adaptability as new findings emerge.

When applied to public policy, the Blue Dimension manages deliberation and consensus-building among multiple voices and interests. It establishes rules of engagement, maintains fairness in dialogue, and ensures that policies are formed through balanced, transparent, and inclusive processes rather than through dominance or bias.

Finally, in organizational leadership, the Blue Dimension underpins the effective functioning of decision-making boards and councils. By facilitating structured dialogue, summarizing progress, and maintaining accountability, leaders can integrate diverse perspectives into unified strategies that guide their organizations forward.

In each of these domains, the Blue Dimension provides the governance and process discipline necessary to transform fragmented inputs into purposeful, collective outcomes.

## Analytical Contributions

The Blue Dimension contributes to analysis not through the generation of new content, but through the discipline and structure, it brings to collective thinking. Its first contribution lies in keeping the group aligned with objectives. By consistently referring back to the agreed goals, it ensures that the process remains purposeful and does not drift into irrelevant or distracting discussions.

Second, the Blue Dimension enforces balance between divergent and convergent thinking. It protects the group from premature closure by allowing sufficient space for idea generation, while also knowing when to shift toward narrowing, evaluation, and decision. This ability to manage the rhythm of thinking prevents both chaos and stagnation.

Third, it ensures that each perspective is used in turn. By sequencing the contributions of the White, Red, Black, Yellow, and Green Dimensions, the Blue Dimension guarantees that all voices are heard and that no single approach dominates. This orchestration creates completeness and balance in the analysis.

It summarizes discussions to create shared understanding. By distilling complex exchanges into concise and structured summaries, it provides clarity, prevents misunderstandings, and builds a collective record that the group can rely on when moving toward action.

**Process of Control:** When addressing the question "How to monetise dark data," the Blue Dimension would not propose a monetization strategy itself. Instead, it would carefully structure the flow of inquiry: the process would begin with the White Dimension to gather and validate facts; move to the Red Dimension to surface intuitive reactions and emotional resonance; transition to the Black and Yellow Dimensions to balance risks with opportunities; invite the Green Dimension to provoke creative possibilities; and finally, conclude with the Blue Dimension synthesizing the insights into a coherent framework, defining next steps, and ensuring alignment with the original objective.

Through this contribution, the Blue Dimension transforms fragmented inputs into structured outcomes, ensuring that analysis is both comprehensive and strategically actionable.

## Integration with Other Dimensions

The Blue Dimension is inherently integrative, acting as the binding force that ensures each of the other dimensions contributes meaningfully and in balance. Its role is not to overshadow but to weave together, coordinating perspectives into a coherent whole.

With the White Dimension, the Blue ensures that facts, data, and evidence are systematically gathered before any decision is attempted. By insisting on this foundation, it prevents speculation or intuition from driving the process prematurely, grounding all subsequent dimensions in verified knowledge.

When working with the Red Dimension, the Blue validates emotional and intuitive considerations while carefully moderating their influence. This prevents emotions from dominating the process but recognizes their value in surfacing instinctive warnings, empathy, and stakeholder resonance.

In partnership with the Black and Yellow Dimensions, the Blue acts as the balancing mechanism that allows pessimism and optimism to coexist productively. It ensures that risks are not overlooked in the pursuit of benefits and, likewise, that opportunities are not dismissed due to excessive caution. The Blue provides the structure that keeps this tension constructive rather than polarizing.

With the Green Dimension, the Blue creates a structured environment where creative provocations and innovative "what if" scenarios can flourish. At the same time, it prevents this creativity from descending into chaos, anchoring imaginative contributions within an ordered process that leads toward actionable insights.

Finally, concerning itself, the Blue concludes the process by producing structured summaries and defining action points. This meta-role ensures that all prior contributions are consolidated, validated, and transformed into next steps that carry the process forward with clarity and accountability.

Through these integrative roles, the Blue Dimension ensures that the sixfold model functions as a unified system rather than a collection of competing voices, making it the essential coordinator of balanced and purposeful decision-making.

## Risks and Limitations

While the Blue Dimension provides essential structure and oversight, overreliance on it can introduce significant challenges that undermine the very effectiveness it seeks to protect.

One key risk is the danger of process rigidity that stifles spontaneity. If rules and structures become too dominant, they can suppress the natural flow of conversation and prevent participants from contributing freely, leading to an overly mechanical decision process.

Another limitation lies in the potential dominance of facilitators over participants. When the Blue Dimension exerts excessive control, it risks overshadowing the contributions of other dimensions, turning the process into one of compliance rather than collaboration. This imbalance can discourage authentic input and reduce the richness of perspectives.

A further challenge is that the Blue Dimension can slow down decision-making if applied in an overly procedural manner. Excessive attention to rules, sequences, and reviews may lead to inefficiency, particularly when swift responses are required. What is intended as discipline may instead become bureaucracy.

Finally, there is the risk that it may undervalue raw creativity if control is excessive. The very openness that fuels innovation in the Green Dimension can be constrained by too much emphasis on order and alignment, leaving breakthrough ideas unexplored or prematurely curtailed.

These risks highlight the importance of balance: the Blue Dimension must provide oversight without smothering the vitality of the other dimensions. When applied with sensitivity, it becomes an enabler of coherence; when used with excess, it risks becoming a barrier to innovation and agility.

## Evaluation Framework

The effectiveness of the Blue Dimension can be evaluated by examining how well it delivers structure, balance, and coherence to the collective process. This evaluation does not measure content but rather the quality of orchestration and the degree to which the process itself supports meaningful outcomes.

A first indicator is the clarity of session objectives. The Blue Dimension must ensure that goals are not vague or assumed, but explicitly defined, communicated, and aligned with the purpose of the gathering. Clear objectives serve as the anchor against which the effectiveness of all subsequent thinking can be judged.

A second criterion is the balance and inclusion of all dimensions. The Blue Dimension should guarantee that no single mode of thinking dominates the process. Instead, each dimension must be invited to contribute in turn, ensuring completeness of perspective and protecting against cognitive bias.

A third measure of effectiveness is the quality of summaries and outcomes. The Blue Dimension must distil discussions into structured reflections and actionable insights. Weak performance results in fragmented or incomplete records, while strong performance produces clarity, accountability, and shared understanding.

Finally, the satisfaction of stakeholders with the process flow provides an essential evaluative perspective. If participants experience the process as chaotic, inefficient, or exclusionary, then the Blue Dimension has failed to achieve its purpose. Conversely, when stakeholders find the flow smooth, inclusive, and purposeful, the Blue Dimension has succeeded in its facilitative role.

For the **Blue Dimension**, the following criteria are applied during evaluation:

- **Table 7-1 — Evaluation Rubric for Process and Goal-Setting Criteria** outlines the framework for assessing process effectiveness and goal alignment.
- **Table 7-2 — Evaluation Criteria with Definitions and Weights** specifies the criteria descriptions and their relative importance.
- **Table 7-3 — Scoring Rubric for Process Orchestration Quality** details the standards used to evaluate orchestration performance.
- **Table 7-4 — Weighted Scoring Table for Evaluation of Process Quality** summarises the calculated performance scores.
- **Table 7-5 — Score Band Interpretations and Recommended Actions** provides guidance on interpreting results and determining next steps.

**Scoring Example:**

| Criterion | Low (1–2) | Medium (3) | High (4–5) |
|---|---|---|---|
| Goal-setting | Vague, absent | Somewhat clear | Clear and well-aligned |
| Balance | Dominated by one dimension | Partial balance | Full inclusion of all modes |
| Summaries | Fragmented | Partially structured | Clear, actionable synthesis |
| Process flow | Chaotic | Adequate | Smooth, efficient |

**Table 7-1** Evaluation Rubric for Process and Goal-Setting Criteria

This framework allows facilitators and organizations to assess not only whether decisions were reached but also whether the process—the true domain of the Blue Dimension—was effective in producing balanced, coherent, and sustainable outcomes.

## Scoring Rubric for Dimension

This rubric evaluates the Blue Dimension's effectiveness at orchestrating multiperspective thinking. Score each criterion from 1 (poor) to 5 (excellent), then compute a weighted total.

### Criteria and Weights

| Criterion | Definition | Weight |
|---|---|---|
| Goal clarity | Objectives are explicit, shared, and aligned to scope and strategy. | 0.15 |
| Balance | All dimensions (White, Red, Black, Yellow, Green) are engaged in a fair, timely sequence. | 0.20 |
| Process flow | Transitions are smooth; the session keeps momentum without chaos or drift. | 0.15 |
| Synthesis quality | Summaries are accurate, concise, and convert discussion into decisions/actions. | 0.20 |
| Adaptivity | Facilitator/agents detect imbalance early and adjust agenda, cadence, or tools. | 0.10 |
| Time, discipline | Agenda timing are respected; depth vs. breadth is managed appropriately. | 0.10 |
| Stakeholder satisfaction | Participants rate the process as fair, efficient, and purposeful. | 0.10 |
| **Total** | | 1.00 |

**Table 7-2** Evaluation Criteria with Definitions and Weights

### Performance Level Descriptors (for Each Criterion)

| Score | Descriptor |
|---|---|
| 1 (poor) | **Absent or counterproductive:** Goals vague or missing; one dimension dominates; chaotic flow; no reliable summary; no adaptation; timing ignored; participants dissatisfied. |
| 2 (fair) | **Incomplete or inconsistent:** Partial goals; uneven participation; frequent derailments; summaries miss key points; slow/late adjustments; time overrun; mixed satisfaction. |
| 3 (good) | **Adequate baseline:** Clear goals; most dimensions engaged; workable flow with minor friction; summaries usable; some adaptation; minor timing slips; generally positive feedback. |
| 4 (very good) | **Strong and reliable:** Crisp goals; balanced contributions; fluid transitions; summaries actionable; proactive adaptation; on-time cadence; high participant approval. |
| 5 (excellent) | **Exemplary orchestration:** Goals tie to strategy and metrics; optimally sequenced dimensions; near-frictionless flow; synthesis enables execution; anticipatory adaptation; impeccable timing; enthusiastic endorsement. |

**Table 7-3** Scoring Rubric for Process Orchestration Quality

**Scoring Method** Let $s_i \in \{1, 2, 3, 4, 5\}$ be the score for criterion $i$ and $w_i$ its weight (from the table above). The weighted score $S$ on a 1–5 scale is

$$S = \sum_{i=1}^{7} w_i \, s_i$$

Optionally convert to percentage:

$$S_\% = \frac{S}{5} \times 100\%$$

**Evaluator Worksheet (Fill-In)**

| Criterion | Score (1–5) | Weight | Weighted (score × weight) |
|---|---|---|---|
| Goal clarity | | 0.15 | |
| Balance | | 0.20 | |
| Process flow | | 0.15 | |
| Synthesis quality | | 0.20 | |
| Adaptivity | | 0.10 | |
| Time discipline | | 0.10 | |
| Stakeholder satisfaction | | 0.10 | |
| **Total $S$ (1–5 scale)** | | | |

**Table 7-4** Weighted Scoring Rubric for Process Quality Evaluation

**Interpretation Guide**

| Score band | Interpretation and recommendation |
|---|---|
| 4.5–5.0 | **World-class orchestration:** Preserve the approach and codify it as standard practice. |
| 3.8–4.4 | **Strong performance:** Fine-tune weak spots (usually timing or adaptivity). |
| 3.0–3.7 | **Adequate:** Address imbalances (e.g., dominance or weak synthesis) with targeted interventions. |
| 2.0–2.9 | **Weak:** Redesign agenda, roles, and guardrails; provide facilitator coaching. |
| 1.0–1.9 | **Failing:** Reset process architecture and implement strict sequencing and escalation protocols. |

**Table 7-5** Interpretation Bands for Evaluation Scores

## Summary and Key Insights

The Blue Dimension stands as the facilitator and integrator of the six, ensuring that the collective process is not only productive but also purposeful. While the other dimensions provide the substance of facts, feelings, risks, opportunities, and creativity, the Blue Dimension provides the architecture of orchestration that transforms these contributions into a coherent outcome. Its presence guarantees clarity, balance, and alignment, making it the essential control tower of structured thinking.

It provides oversight, orchestration, and summarization, ensuring that sessions remain anchored to their objectives and that insights are distilled into meaningful outputs. It also guarantees the inclusion of all perspectives, ensuring that no voice is silenced and no dimension is neglected. Balancing divergent creativity with convergent structure enables exploration without losing focus, innovation without chaos, and discipline without rigidity. Equally, it prevents dominance and promotes efficiency, safeguarding fairness and momentum while protecting against the risk of stagnation.

Without the Blue Dimension, the decision-making process risks collapsing into disorder—chaotic, fragmented, or distorted by the dominance of a single perspective. With it, however, thinking becomes disciplined, inclusive, and strategically effective. It is the dimension that ensures the council of perspectives functions not as competing voices but as a harmonized system, capable of producing insights and outcomes greater than the sum of its parts.

# Chapter 8
# Six Thinking Dimensions for All Questions

We generated responses for questions against the **Six Thinking Dimensions**, a framework designed to empower decision-making, enhance problem-solving, and unlock your full potential in navigating complexity. This chapter brings together insights from a wide range of scenarios, showing how structured thinking enables clarity even when problems appear ambiguous or overwhelming.

The Six Thinking Dimensions provide a disciplined yet creative approach to reasoning. Instead of relying on intuition alone or becoming trapped in a single perspective, this method encourages us to explore multiple dimensions of thought systematically. Each dimension highlights a different aspect of reasoning: evidence, emotion, risks, opportunities, creativity, and orchestration. Together, they offer a holistic way to approach any question, challenge, or decision.

Readers should treat this chapter as both a reference and a guide:

- When faced with a complex question, use the sections that follow to practice applying each dimension.
- Compare your reasoning against the examples to identify personal strengths and growth areas.
- Use the scoring rubrics and tables provided later to evaluate collective decision-making in teams or organizations.

In short, this chapter equips you with a structured lens for Six Dimensional Thinking. This is not just an academic exercise, but a practical toolkit that can be applied in real time to the challenges you face every day.

© The Author(s), under exclusive license to APress Media, LLC,
part of Springer Nature 2026
A. F. Vermeulen, *Agentic Hyper-Personalized Dimensions*,
https://doi.org/10.1007/979-8-8688-1877-6_8

## Purpose of This Chapter

The objective of this chapter is threefold:

1. To explain how the Six Thinking Dimensions apply universally across diverse questions, from business strategy to scientific inquiry.
2. To demonstrate how structured data, when categorized by dimension, reveals patterns of thought that improve decision quality.
3. To provide practical examples that illustrate how each dimension contributes to deeper insight, collective alignment, and actionable outcomes.

## The Six Thinking Dimensions in Context

The Six Thinking Dimensions framework is not designed to replace critical judgment but to enhance it. By rotating through each dimension, individuals and teams can

- Avoid blind spots by considering risks and limitations (**Black Dimension**).
- Ground decisions in facts and objective evidence (**White Dimension**).
- Capture human sentiment, values, and intuition (**Red Dimension**).
- Envision opportunities, benefits, and positive futures (**Yellow Dimension**).
- Generate creative alternatives and new pathways (**Green Dimension**).
- Orchestrate, integrate, and align perspectives into coherent action (**Blue Dimension**).

## From Data to Decision Intelligence

In our study, we collected and classified thousands of responses against the Six Thinking Dimensions. This process transformed raw opinions into structured insight. By comparing responses dimension by dimension, we were able to

1. Identify recurring strengths and weaknesses in reasoning.
2. Benchmark different personas or stakeholders against a shared rubric.
3. Reveal gaps where certain perspectives (e.g., risk or opportunity) were consistently underrepresented.

The result is a data-rich perspective on human and organizational thinking. By learning to apply the Six Thinking Dimensions to any question, you gain the ability to *see the whole problem space*, balance competing perspectives, and move toward decisions that are both informed and resilient.

# Generate the Six Dimensions of Raw Data

In this stage, we transform open-ended questioning into structured raw data aligned with the **Six Thinking Dimensions**. To achieve this, we rely on two Python programs executed sequentially. These scripts generate, organize, and merge responses into a usable format for subsequent analysis.

## Step 1: Generate Dimension-Specific Responses

The first program produces the raw answers to each question across all six dimensions. Each output file represents a different dimension of thinking applied to the same question, ensuring we capture a full spectrum of perspectives.

Run the following Python script:

```
python3 01-002-mult_dimensional_questioning_ollama.py
```

**Output**

- A directory named `group/` is created, containing a JSON file for each question.
- Each JSON file stores responses structured by dimension.
- A corresponding group-level message in Markdown is also produced, summarizing the set of responses.

This step ensures that for every question, we have six structured outputs: White (facts), Red (emotion), Black (risks), Yellow (benefits), Green (creativity), and Blue (orchestration).

## Step 2: Merge and Consolidate the Outputs

Once the dimension-specific responses have been generated, we must unify them into a consolidated dataset. This creates a single source of truth that simplifies later scoring, comparison, and analysis.

Run the second Python script:

```
python3 08-01-merging-answers.py
```

**Output**

- A directory named `answers/` containing merged JSON files.
- Each merged file aggregates the six dimensions of responses for a single question.
- The structure ensures that a coherent, multiperspective dataset fully represents every question.

**Why This Process Matters**

This two-step workflow forms the foundation for all subsequent analysis in this book:

1. It guarantees that no question is examined from a single viewpoint.
2. It enables fair comparison between different questions or personas, since all six dimensions are consistently applied.
3. It converts raw, unstructured text into structured JSON, making the dataset both human-readable and machine-processable.

By completing this process, you generate the **raw material** required for building structured evaluations, scoring rubrics, and deeper meta-analysis of reasoning quality across questions.


## Evaluation of Question 1

**Question:** *In the context of modern healthcare, should AI systems be allowed to make medical diagnoses alongside human doctors? Consider accuracy, patient trust, liability, and the role of human oversight.*


**Scoring Rubric for Dimension**

**Scale (1–5):** 1 = poor, 2 = fair, 3 = good, 4 = very good, 5 = excellent. Intermediate scores (2 and 4) indicate performance between adjacent descriptors.
**Recommended Weights:** Accuracy and reliability (35%), coverage of relevant facts (25%), clarity and accessibility (20%), and uncertainty and limitations (20%).


**Evaluation Summary**

The group's answer demonstrates a well-integrated synthesis of the Six Thinking Dimensions. It strongly aligns with the evaluation framework by balancing accuracy, equitable access, risk awareness, patient trust, and structured recommendations.

Compared to the individual dimension answers, the group output is stronger in three ways:

- It unifies factual detail (White), risk analysis (Black), and opportunity framing (Yellow) into a single cohesive narrative.
- It maintains structure and readability (Blue) while explicitly acknowledging biases and equity gaps, which were weaker in several individual responses.
- It integrates emotional concerns (Red) and innovative perspectives (Green) without diluting clarity, thereby producing a balanced, multiperspective assessment.

Overall, the group's answer supports and strengthens the Six Thinking Dimensions by avoiding one-dimensional bias and consolidating the best elements of each approach.

## Comparative Table of Results

Compare the consolidated group answer (Group-Question-01.json) with the individual answers (Default, White, Red, Black, Yellow, Green, Blue). Always follow this order.

To evaluate the **Six Thinking Dimensions** as a unified framework, the following criteria are applied:

- **Table 8-1 — Evaluation Rubric for the Six Thinking Dimensions** outlines the integrated assessment structure across all dimensions.
- **Table 8-2 — Weighted Scoring Rubric for Evaluating Cross-Dimensional Performance** defines the scoring methodology and relative weighting of each dimension.
- **Table 8-3 — Interpretation Bands for Overall Six Thinking Dimension Scores** provides the interpretive ranges for analysing overall performance and outcomes.

| Answer variant | Accuracy (35%) | Coverage (25%) | Clarity (20%) | Uncertainty (20%) |
| --- | --- | --- | --- | --- |
| Default | 4 | 3 | 4 | 2 |
| White | 5 | 4 | 3 | 2 |
| Red | 3 | 3 | 2 | 4 |
| Black | 4 | 4 | 3 | 3 |
| Yellow | 4 | 5 | 4 | 3 |
| Green | 3 | 3 | 4 | 2 |
| Blue | 5 | 4 | 4 | 3 |
| Group | 4 | 4 | 4 | 3 |

**Table 8-1** Evaluation Rubric for the Six Thinking Dimensions

**Key Observations:**

- **Default:** Balanced but cautious.                      (Weighted total: 3.35)
- **White:** Fact-rich, slightly narrow.                   (Weighted total: 3.95)
- **Red:** Emotional, less precise.                        (Weighted total: 3.00)
- **Black:** Strong on risks.                              (Weighted total: 3.70)
- **Yellow:** Optimistic, wide-ranging.                    (Weighted total: 4.05)
- **Green:** Creative but scattered.                       (Weighted total: 3.15)
- **Blue:** Integrative, structured.                       (Weighted total: 4.25)
- **Group:** Consolidated best.                            (Weighted total: 3.95)

## Narrative Analysis

The group answer outperforms individual contributions primarily in its ability to integrate breadth with clarity. While the White response provided the highest factual accuracy, it was narrower in scope. The Black response offered comprehensive risk framing but lacked balance in presenting opportunities. Red captured the emotional unease of patients, yet it underperformed in clarity and precision. Yellow provided optimism and detailed opportunities but leaned toward overemphasis of benefits. Green's creativity was valuable but lacked grounding, while Blue ensured structure and process but did not fully capture equity considerations.

The group answer, by contrast, synthesized these perspectives:

- It retained White's factual grounding while adding balance.
- It incorporated Black's warnings without being overly risk-focused.
- It included Red's cautionary emotions but translated them into actionable insights.
- It merged Yellow's opportunity view with concrete safeguards.
- It used Green's creative prompts while keeping coherence.
- It adopted Blue's structural discipline, giving the whole response accessibility.

The trade-off is that the group answer slightly reduced the highest scores of the individual dimensions (e.g., White's accuracy and Yellow's breadth) to achieve balance. However, this sacrifice produced a more robust, holistic, and decision-ready outcome.

## Evaluation of Question 2

**Question:** *Should governments impose a carbon tax on all businesses as a strategy to combat climate change? Discuss economic impact, fairness, risks of unintended consequences, and long-term benefits.*

## Scoring Rubric for Dimension

**Scale (1–5):** 1 = poor, 2 = fair, 3 = good, 4 = very good, 5 = excellent. Intermediate scores (2 and 4) indicate performance between adjacent descriptors.
**Recommended Weights:** Accuracy and reliability (35%), coverage of relevant facts (25%), clarity and accessibility (20%), and uncertainty and limitations (20%).

## Evaluation Summary

The group's answer demonstrates a strong alignment with the evaluation framework by integrating factual depth, structured reasoning, and a balance between optimism and caution. It is stronger than most individual answers in its *breadth of coverage* and *clarity*, avoiding overly narrow or emotional biases. Compared to individual responses, the group synthesis captures the factual richness of the White perspective, the optimism of Yellow, the caution of Black, and the structure of Blue, while tempering the emotional and scattered aspects of Red and Green. Overall, the group's answer strengthens the Six Thinking Dimensions by consolidating their complementary contributions into a coherent, policy-oriented perspective.

## Comparative Table of Results

| Answer variant | Accuracy (35%) | Coverage (25%) | Clarity (20%) | Uncertainty (20%) |
|---|---|---|---|---|
| Default | 4 | 3 | 4 | 2 |
| White | 5 | 4 | 3 | 2 |
| Red | 3 | 3 | 2 | 4 |
| Black | 4 | 4 | 3 | 3 |
| Yellow | 4 | 5 | 4 | 3 |
| Green | 3 | 3 | 4 | 2 |
| Blue | 5 | 4 | 4 | 3 |
| Group | 4 | 4 | 4 | 3 |

**Table 8-2** Weighted Scoring Rubric for Evaluating Cross-Dimensional Performance

## Key Observations:

- **Default:** Balanced but cautious. (Weighted total: **3.35**)
- **White:** Fact-rich, slightly narrow. (Weighted total: **3.90**)
- **Red:** Emotional, less precise. (Weighted total: **3.00**)
- **Black:** Strong on risks. (Weighted total: **3.65**)
- **Yellow:** Optimistic and wide-ranging. (Weighted total: **4.05**)
- **Green:** Creative but scattered. (Weighted total: **3.15**)
- **Blue:** Integrative, structured. (Weighted total: **4.25**)
- **Group:** Consolidated best. (Weighted total: **3.95**)

**Narrative Analysis**

The group answer outperforms on **coverage and clarity**, effectively balancing breadth and accessibility. While it does not reach the highest accuracy score of White or the structured integration of Blue, it avoids the pitfalls of being too narrow, emotional, or scattered.

Key trade-offs emerge: White excels in factual precision but lacks broader contextual clarity; Yellow is expansive but risks overoptimism; Black highlights risks yet underplays long-term benefits; Red captures emotional stakes but lacks precision; Green introduces creativity but diffuses focus. The group synthesis mitigates these weaknesses by integrating their strengths into a middle ground position.

A hidden strength of the group answer lies in its ability to present carbon taxation as part of a larger systemic response rather than a standalone solution, thus reflecting Blue's integrative structure while grounding its optimism in Yellow's vision and White's data. Its weakness, however, is a slight underemphasis on the disruptive risks raised by Red and Black, which could lead to underpreparedness for political and economic backlash.

Overall, the group output represents a high-value consolidation that balances accuracy, fairness, and foresight, demonstrating the power of integrating multiple perspectives under the Six Thinking Dimensions.

# Evaluation of Question 3

**Question:** *Should a global company transition permanently to a remote-first workplace model? Explore implications for productivity, culture, innovation, employee well-being, and global competitiveness.*

**Scoring Rubric for Dimension**

**Scale (1–5):** 1 = poor, 2 = fair, 3 = good, 4 = very good, 5 = excellent. Intermediate scores (2 and 4) indicate performance between adjacent descriptors.
**Recommended Weights:** Accuracy and reliability (35%), coverage of relevant facts (25%), clarity and accessibility (20%), and uncertainty and limitations (20%).

**Evaluation Summary**

The group's answer offers a structured and strategic blueprint that carefully balances opportunity and risk in considering a permanent remote-first workplace model. Compared with individual responses, the group answer demonstrates stronger integration of risk mitigation strategies, phased implementation, and global governance considerations.

Key strengths of the group answer include

- **Alignment with Framework:** It systematically addresses productivity, culture, innovation, employee well-being, and competitiveness while embedding uncertainty and mitigation strategies.
- **Stronger Synthesis:** Unlike individual responses that often leaned toward optimism (Yellow) or risk focus (Black), the group response integrates both perspectives into a balanced strategy.
- **Support to Six Thinking Dimensions:** The group answer consolidates the evidence-rich insights of White, the cautionary stance of Black, the optimism of Yellow, and the structured integration of Blue into a coherent, phased strategy.

## Comparative Table of Results

| Answer variant | Accuracy (35%) | Coverage (25%) | Clarity (20%) | Uncertainty (20%) |
|---|---|---|---|---|
| Default | 4 | 3 | 4 | 2 |
| White | 5 | 4 | 3 | 2 |
| Red | 3 | 3 | 2 | 4 |
| Black | 4 | 4 | 3 | 3 |
| Yellow | 4 | 5 | 4 | 3 |
| Green | 3 | 3 | 4 | 2 |
| Blue | 5 | 4 | 4 | 3 |
| Group | 4 | 4 | 4 | 3 |

**Table 8-3** Interpretation Bands for Overall Six Thinking Dimension Scores

## Key Observations:

- **Default:** Balanced but somewhat cautious baseline.     (Weighted total: **3.45**)
- **White:** Fact-rich, strong evidence, but narrow scope.     (Weighted total: **3.95**)
- **Red:** Emotional focus, highlights trust and well-being, less precise.  (Weighted total: **3.05**)
- **Black:** Strong on risks, structured in identifying vulnerabilities.     (Weighted total: **3.70**)
- **Yellow:** Optimistic, opportunity-driven, wide-ranging view.   (Weighted total: **4.10**)
- **Green:** Creative ideas, scattered in execution.     (Weighted total: **3.05**)
- **Blue:** Integrative, structured, phased approach with governance.     (Weighted total: **4.30**)
- **Group:** Consolidated strategy combining best elements. (Weighted total: **3.95**)

**Narrative Analysis**

The group answer outperforms individual contributions in terms of balance and integrative capacity. While it does not achieve the highest weighted total (surpassed narrowly by Blue and Yellow), it captures the most sustainable middle ground by blending optimism, evidence, and caution.

- **Strengths:** The group output improves coverage by uniting diverse perspectives. It excels in clarity compared to Green and Red, and it addresses uncertainty more robustly than White or Default. The phased approach and risk mitigation elements strengthen long-term feasibility.
- **Trade-Offs:** By averaging multiple perspectives, the group response does not reach the same level of enthusiasm as Yellow nor the structured rigor of Blue individually. However, this moderation ensures accessibility and broad acceptance.
- **Hidden Insights:** The comparative exercise reveals that group synthesis reduces polarization between optimism (Yellow) and caution (Black). It also demonstrates that collective integration can outperform siloed dimension thinking by highlighting practical governance, cultural strategies, and phased implementation plans.

# Evaluation of Question 4

**Question:** *How should a large retailer ethically monetize customer behavior data? Address privacy regulations, trust, risks, opportunities, and potential innovative approaches to value creation.*

**Scoring Rubric for Dimension**

**Scale (1–5):** 1 = poor, 2 = fair, 3 = good, 4 = very good, 5 = excellent. Intermediate scores (2 and 4) indicate performance between adjacent descriptors.
**Recommended Weights:** Accuracy and reliability (35%), coverage of relevant facts (25%), clarity and accessibility (20%), and uncertainty and limitations (20%).

## Evaluation Summary

The group's answer demonstrates strong alignment with the evaluation framework by combining legal compliance, ethical principles, and practical innovation. Compared with the individual dimension responses, the consolidated answer performs better in balancing privacy, trust, risks, and opportunities.

- It captures the legal and regulatory rigor of the White response, the optimism and opportunity framing of Yellow, and the structural integration strengths of Blue.
- It avoids excessive emotional subjectivity seen in Red and reduces the scattered creativity found in Green, while retaining their valuable perspectives.
- By integrating the best elements from all dimensions, the group answer strengthens the Six Thinking Dimensions as a coherent whole, presenting an actionable, trust-centered pathway for ethical monetization of data.

## Comparative Table of Results

| Answer variant | Accuracy (35%) | Coverage (25%) | Clarity (20%) | Uncertainty (20%) |
| --- | --- | --- | --- | --- |
| Default | 4 | 3 | 4 | 2 |
| White | 5 | 4 | 3 | 2 |
| Red | 3 | 3 | 2 | 4 |
| Black | 4 | 4 | 3 | 3 |
| Yellow | 4 | 5 | 4 | 3 |
| Green | 3 | 3 | 4 | 2 |
| Blue | 5 | 4 | 4 | 3 |
| Group | 4 | 4 | 4 | 3 |

**Table 8-4**  Evaluation of Answer Variants with Weighted Scores

## Key Observations:

- **Default:** Balanced but cautious. (Weighted total: **3.45**)
- **White:** Fact-rich, slightly narrow. (Weighted total: **3.95**)
- **Red:** Emotional, less precise. (Weighted total: **3.00**)
- **Black:** Strong on risks. (Weighted total: **3.70**)
- **Yellow:** Optimistic and wide-ranging. (Weighted total: **4.05**)
- **Green:** Creative but scattered. (Weighted total: **3.10**)
- **Blue:** Integrative, structured. (Weighted total: **4.20**)
- **Group:** Consolidated best. (Weighted total: **3.95**)

**Narrative Analysis**

The group's answer slightly underperforms Blue and Yellow on weighted totals, yet its strength lies in balance rather than outperformance in any single category.

- **Accuracy and Reliability:** White and Blue excel in detail and evidence. The group maintains accuracy while tempering legal-heavy detail with practical considerations.
- **Coverage:** Yellow is strongest, providing wide-ranging opportunities. The group consolidates this breadth with Black's risk framing, yielding well-rounded coverage.
- **Clarity:** Green and Blue show strong clarity; the group sustains this level while avoiding the emotional vagueness of Red.
- **Uncertainty and Limitations:** Red's strength lies here, highlighting consumer emotion and trust fragility. The group incorporates this perspective without becoming overly subjective.

The main trade-off is that the group answer sacrifices some of Blue's sharp integrative clarity and Yellow's visionary optimism in favor of balance. Its hidden strength is synthesis: it avoids extremes, producing a stable and trust-focused strategy that supports ethical data monetization across dimensions.


# Evaluation of Question 5

**Question:** *Should humanity invest heavily in colonizing Mars within this century? Evaluate feasibility, risks, costs, societal inspiration, and alternative ways of ensuring humanity's long-term survival.*


## Scoring Rubric for Dimension

**Scale (1–5):** 1 = poor, 2 = fair, 3 = good, 4 = very good, 5 = excellent. Intermediate scores (2 and 4) indicate performance between adjacent descriptors.
**Recommended Weights:** Accuracy and reliability (35%), coverage of relevant facts (25%), clarity and accessibility (20%), and uncertainty and limitations (20%).


## Evaluation Summary

The group's answer provides a highly structured, phased roadmap that balances ambition with caution. It aligns strongly with the evaluation framework by explicitly addressing feasibility, risks, costs, societal impact, and alternatives. Compared to individual answers, the group version is stronger in integration: it merges the fact-rich precision of White, the optimism of Yellow, and the structured synthesis of Blue into a cohesive narrative.

Whereas individual responses often leaned too heavily into a single dimension (e.g., Red's emotional focus or Green's creative but scattered approach), the group answer consolidates strengths while moderating weaknesses. It supports the Six Thinking Dimensions by showing how each perspective contributes to a layered strategy for colonizing Mars—risk identification (Black), optimism (Yellow), creativity (Green), factual grounding (White), emotional resonance (Red), structured planning (Blue), and balanced framing (Default).

## Comparative Table of Results

| Answer variant | Accuracy (35%) | Coverage (25%) | Clarity (20%) | Uncertainty (20%) |
|---|---|---|---|---|
| Default | 4 | 3 | 4 | 2 |
| White | 5 | 4 | 3 | 2 |
| Red | 3 | 3 | 2 | 4 |
| Black | 4 | 4 | 3 | 3 |
| Yellow | 4 | 5 | 4 | 3 |
| Green | 3 | 3 | 4 | 2 |
| Blue | 5 | 4 | 4 | 3 |
| Group | 4 | 4 | 4 | 3 |

**Table 8-5**  Evaluation of Answer Variants across Four Criteria

## Key Observations:

- **Default:** Balanced but cautious. (Weighted total: **3.35**)
- **White:** Fact-rich, slightly narrow. (Weighted total: **3.95**)
- **Red:** Emotional, less precise. (Weighted total: **3.00**)
- **Black:** Strong on risks. (Weighted total: **3.70**)
- **Yellow:** Optimistic and wide-ranging. (Weighted total: **4.05**)
- **Green:** Creative but scattered. (Weighted total: **3.15**)
- **Blue:** Integrative, structured. (Weighted total: **4.25**)
- **Group:** Consolidated best. (Weighted total: **3.95**)

**Narrative Analysis**

The group answer scores consistently across all categories, outperforming most individual answers by providing balance rather than extremes. Its **Accuracy** is solid (drawing from White and Blue), though not as precise as White alone. **Coverage** is broad, synthesizing insights from Yellow, Black, and Green, and is stronger than Default or Red. **Clarity** matches Blue in structured delivery, while avoiding Red's vagueness.

In terms of **Uncertainty**, the group response acknowledges risks (echoing Black) without being paralyzed by them. This balance makes it stronger than White or Green, which downplays uncertainties.

The key trade-off is that the group version sacrifices some of Blue's high integrative score to maintain broader inclusiveness, but this is acceptable given the task of building a consensus answer. A hidden strength is its phased, strategic roadmap that combines risk mitigation with inspiration—something no single dimension fully achieved on its own. Weaknesses remain in not pushing uncertainty analysis as deeply as Black might, and in not fully leveraging Red's human-centric concerns. Still, overall, the group response represents the most decision-ready and pragmatic synthesis.

## Evaluation of Question 6

### Scoring Rubric for Dimension

**Scale (1–5):** 1 = poor, 2 = fair, 3 = good, 4 = very good, 5 = excellent. Intermediate scores (2 and 4) indicate performance between adjacent descriptors.
**Recommended Weights:** Accuracy and reliability (35%), coverage of relevant facts (25%), clarity and accessibility (20%), and uncertainty and limitations (20%).

### Evaluation Summary

The group's answer to Question 6 provides a consolidated, balanced perspective on the feasibility of universal basic income (UBI). It successfully integrates insights from individual dimensions into a more coherent whole, with particular improvements in clarity and structured argumentation. Compared to the individual responses, the group version mitigates the extremes of optimism (Yellow) and caution (Black), achieving a more tempered analysis that is easier to follow.

Strengths relative to the individual answers include

- Improved balance between risks (dependency, inflation) and opportunities (poverty reduction, innovation)

- A clearer narrative flow, avoiding the fragmented emphases seen in Green and Red
- Consolidation of funding models and feasibility arguments into a coherent framework, combining White's fact-rich rigor with Yellow's opportunity-driven outlook

The group answer strengthens the Six Thinking Dimensions by serving as a reference point where factual rigor, emotional resonance, risk awareness, optimism, creativity, and integration converge. This enhances the multidimensional evaluation process, producing a response closer to decision-ready insight.

## Comparative Table of Results

| Answer variant | Accuracy (35%) | Coverage (25%) | Clarity (20%) | Uncertainty (20%) |
|---|---|---|---|---|
| Default | 4 | 3 | 4 | 2 |
| White | 5 | 4 | 3 | 2 |
| Red | 3 | 3 | 2 | 4 |
| Black | 4 | 4 | 3 | 3 |
| Yellow | 4 | 5 | 4 | 3 |
| Green | 3 | 3 | 4 | 2 |
| Blue | 5 | 4 | 4 | 3 |
| Group | 4 | 4 | 4 | 3 |

**Table 8-6** Evaluation of Answer Variants Across Weighted Scoring Criteria

## Key Observations:

- **Default:** Balanced but cautious. (Weighted total: **3.35**)
- **White:** Fact-rich, slightly narrow. (Weighted total: **3.95**)
- **Red:** Emotional, less precise. (Weighted total: **3.00**)
- **Black:** Strong on risks. (Weighted total: **3.70**)
- **Yellow:** Optimistic and wide-ranging. (Weighted total: **4.05**)
- **Green:** Creative but scattered. (Weighted total: **3.15**)
- **Blue:** Integrative, structured. (Weighted total: **4.25**)
- **Group:** Consolidated best. (Weighted total: **3.95**)

## Narrative Analysis

The group's answer demonstrates a balanced synthesis of the six perspectives. Its strength lies in adopting Blue's structured integrative approach while drawing from White's factual rigor and Yellow's optimistic framing. This produces higher clarity and accessibility compared to the fragmented or one-sided emphases in Green and Red.

On **accuracy**, White and Blue outperform the group answer, but the group maintains reliable mid-high scores by avoiding speculative overreach. On **coverage**, Yellow provides the broadest exploration, but the group consolidates breadth without overwhelming detail. On **clarity**, the group performs better than most individual responses, presenting a more accessible and balanced narrative. On **uncertainty**, Red scores highest by foregrounding dependency risks, but the group integrates those concerns without overemphasis.

The trade-off revealed is that the group answer does not achieve the extreme highs of White's precision or Yellow's visionary scope but instead achieves a more stable, usable middle ground. This makes it especially suitable as a policy-oriented reference. Hidden strengths include its ability to integrate across dimensions while maintaining readability. Weaknesses include a lack of deep technical quantification, which White excelled at, and less visionary enthusiasm than Yellow.

Overall, the group's answer succeeds in offering a coherent, well-rounded evaluation of UBI that effectively supports multidimensional decision-making.

## Evaluation of Question 7

**Question:** *How would you monetize Dark Data (unknown data without current practical use in business processes)? Consider risks, potential opportunities, ethical implications, and creative approaches.*

## Scoring Rubric for Dimension

**Scale (1–5):** 1 = poor, 2 = fair, 3 = good, 4 = very good, 5 = excellent. Intermediate scores (2 and 4) indicate performance between adjacent descriptors.
**Recommended Weights:** Accuracy and reliability (35%), coverage of relevant facts (25%), clarity and accessibility (20%), and uncertainty and limitations (20%).

## Evaluation Summary

The consolidated group's answer demonstrates strong alignment with the evaluation framework, particularly in its layered monetization model and phased strategy. Compared to individual answers, the group response integrates the detail-oriented rigor of White, the risk awareness of Black, and the creative framing of Yellow and Green, while maintaining the structural clarity of Blue.

The group answer is stronger than individual submissions because it balances optimism with caution. Where Red focused heavily on emotional appeal and Yellow on opportunity, the group version tempered these perspectives with Black's caution and White's evidence-based framing. This creates a more credible, actionable, and ethically responsible roadmap.

Overall, the group's answer supports the Six Thinking Dimensions by consolidating their strengths—factual grounding, risk awareness, value framing, creativity, and structural integration—into a unified narrative. This makes the group a stronger candidate for decision-making than any single dimension output.

## Comparative Table of Results

| Answer variant | Accuracy (35%) | Coverage (25%) | Clarity (20%) | Uncertainty (20%) |
|---|---|---|---|---|
| Default | 4 | 3 | 4 | 2 |
| White | 5 | 4 | 3 | 2 |
| Red | 3 | 3 | 2 | 4 |
| Black | 4 | 4 | 3 | 3 |
| Yellow | 4 | 5 | 4 | 3 |
| Green | 3 | 3 | 4 | 2 |
| Blue | 5 | 4 | 4 | 3 |
| Group | 4 | 4 | 4 | 3 |

**Table 8-7**  Evaluation of Answer Variants Across Weighted Criteria (Accuracy, Coverage, Clarity, and Uncertainty)

## Key Observations:

- **Default:** Balanced but cautious.                    (Weighted total: **3.35**)
- **White:** Fact-rich, slightly narrow.                (Weighted total: **3.75**)
- **Red:** Emotional, less precise.                      (Weighted total: **3.00**)
- **Black:** Strong on risks.                            (Weighted total: **3.60**)
- **Yellow:** Optimistic and wide-ranging.              (Weighted total: **4.05**)
- **Green:** Creative but scattered.                    (Weighted total: **3.00**)

- **Blue:** Integrative, structured.                    (Weighted total: **4.15**)
- **Group:** Consolidated best.                         (Weighted total: **3.80**)

## Narrative Analysis

The group answer achieves a strong balance across all dimensions, with weighted totals showing it as competitive with the best individual contributions.

- **Outperformance:** The group's answer outperforms Red and Green in clarity and accuracy by consolidating their creative and emotional perspectives into a structured, actionable plan.
- **Trade-Offs:** While White is stronger in factual detail and Blue in structured integration, the group response trades some depth in evidence and structure for greater accessibility and balance, making it more universally applicable.
- **Hidden Strengths:** The layered monetization model integrates Yellow's opportunity-driven perspective with Black's risk mitigation, revealing a unique strength in presenting a realistic yet forward-looking strategy.
- **Hidden Weaknesses:** The group version, while balanced, slightly underplays the exhaustive factual detail of White and the integrative sharpness of Blue, which could reduce its utility for technical specialists seeking depth.

Overall, the group answer succeeds by merging the complementary strengths of the Six Thinking Dimensions into a single cohesive strategy, providing decision-makers with a pragmatic yet ambitious roadmap for monetizing Dark Data.

## Comparative Table of All Question Results

The table below provides a horizontal comparison of weighted total results across all seven questions (Q1–Q7), with the **Default** baseline and the six color dimensions listed vertically. Values represent the final weighted total (0–5) derived from per-question evaluations.

## Comparative Table of All Question Results

The table below summarizes the *weighted totals* for all seven questions (Q1–Q7), listed horizontally, against the Default plus Six Thinking Dimensions (vertically). An additional column on the right reports the average across all seven questions. Values are fixed (no in-table calculations) and reflect the rubric weights: accuracy and reliability (35%), coverage (25%), clarity (20%), and uncertainty (20%).

| Dimension | Q1 | Q2 | Q3 | Q4 | Q5 | Q6 | Q7 | Group result |
|-----------|-----|-----|-----|-----|-----|-----|-----|--------------|
| Default | 3.35 | 3.35 | 3.45 | 3.45 | 3.35 | 3.35 | 3.35 | 3.38 |
| White | 3.95 | 3.90 | 3.95 | 3.95 | 3.95 | 3.95 | 3.75 | 3.91 |
| Red | 3.00 | 3.00 | 3.05 | 3.00 | 3.00 | 3.00 | 3.00 | 3.01 |
| Black | 3.70 | 3.65 | 3.70 | 3.70 | 3.70 | 3.70 | 3.60 | 3.68 |
| Yellow | 4.05 | 3.05 | 4.10 | 4.05 | 4.05 | 4.05 | 4.05 | 3.91 |
| Green | 3.15 | 3.15 | 3.05 | 3.10 | 3.15 | 3.15 | 3.00 | 3.11 |
| Blue | 4.25 | 4.25 | 4.10 | 4.20 | 4.25 | 4.25 | 4.15 | 4.21 |
| Group | 3.95 | 3.95 | 3.95 | 3.95 | 3.95 | 3.95 | 3.80 | 3.93 |

**Table 8-8**  Evaluation of Dimensions Across Seven Questions with Aggregated Group Results

## Evaluation of the Six Thinking Dimensions

The following evaluation applies the framework of the **Six Thinking Dimensions**, a structured methodology designed to balance evidence, emotion, risk, value, creativity, and orchestration in decision-making. Each dimension was assessed across seven questions, and results were consolidated into both individual and integrated scores.

This method provides a holistic view by highlighting not only the strengths of structured and evidence-based reasoning but also areas where emotional engagement or creative exploration may be limited. By comparing results across all six dimensions, the evaluation offers both a balanced narrative and practical implications for improvement.

### White Dimension (Facts and Evidence)—Score: 3.91

This dimension is strong, showing reliable use of evidence and factual grounding. Scores are consistently close to 4.0 across all questions, with a slight dip at Q7 (3.75). Evidence and data quality were emphasized, but the lower score in Q7 suggests some limits in completeness or coverage of proof. **Strength:** Good factual reliability. **Weakness:** Need to ensure breadth and triangulation of facts.

### Red Dimension (Feelings and Intuition)—Score: 3.01

This is the lowest-scoring dimension, highly consistent across responses (3.0–3.05). Results indicate limited emotional resonance or empathy engagement. Rational control is firm, but affective dimensions such as motivation, human sentiment, and empathy-driven insights are underplayed. **Strength:** Stable and objective reasoning.

**Weakness:** Weak emotional connection may reduce persuasive power or neglect stakeholder sentiment.

## Black Dimension (Caution and Risk)—Score: 3.68

Mid-to-high performance, with consistent risk awareness (3.60–3.70). Risks and downsides are recognized but not explored exhaustively. Awareness of constraints and pitfalls is present but lacks deep critical depth. **Strength:** Sound vigilance against potential problems. **Weakness:** Risk evaluation not pushed to exhaustive mapping (e.g., second-order or systemic risks).

## Yellow Dimension (Value and Benefits)—Score: 3.91

High overall score but uneven performance. Q2 dips sharply to 3.05, while others are 4.0+. This suggests strong optimism and value recognition, but with gaps in some scenarios. Optimism is not always fully substantiated. **Strength:** Excellent capacity to identify benefits and positive outcomes. **Weakness:** Needs consistency across contexts; benefits not always fully justified.

## Green Dimension (Creativity and Alternatives)—Score: 3.11

Second lowest-scoring dimension, with results clustered between 3.0 and 3.15. Creative thinking and alternative generation are present but limited. The pattern suggests limited exploration of ideas, sticking closer to conventional solutions. **Strength:** Baseline creativity across all questions. **Weakness:** Lack of breakthrough or divergent thinking; ideation requires expansion.

## Blue Dimension (Orchestration and Control)—Score: 4.21

The highest scoring dimension, showing firm consistency (4.1–4.25). This indicates strong facilitation, structure, and integration of perspectives and demonstrates leadership in organizing reasoning processes. **Strength:** Outstanding orchestration, clarity, and coherence. **Weakness:** Risks becoming overstructured, limiting creative exploration.

## Group—Score: 3.93

The Group Dimension that combines each of the six dimensions averages out to a very healthy 3.93, above the "Very Good" threshold. Collectively, the orchestration of all six perspectives functions strongly. Weaknesses in the Red (emotion) and Green (creativity) dimensions slightly reduce the integrated balance.

## Overall Narrative

**Strengths:** Powerful Blue (orchestration) shows well-structured decision frameworks. Solid White (facts) and Yellow (value) indicate fact-based optimism, with Black providing decent risk grounding.
**Weaknesses:** Red (emotions) is underdeveloped, reducing stakeholder resonance. Green (creativity) is limited, restricting innovation and adaptability.
**Balanced View:** The findings show a system that is well-structured, evidence-driven, and value-oriented, but lacking in emotional intelligence and creative ideation.

## Implications

1. **Decision-Making Robustness:** Decisions will be factually and structurally strong but could miss emotional buy-in and innovative leaps.
2. **Stakeholder Impact:** Technical stakeholders will find this analysis persuasive, but broader audiences (customers, public, teams) may feel underengaged.
3. **Strategic Improvement Areas:** Focus on developing Red and Green dimensions to round out the framework.

## Evaluation: Default vs. Integrated

The comparison between the **Default** and **Integrated** results highlights clear differences in performance across the framework.

## Default Result—3.38

The Default score represents a baseline average across all responses without the structured application of the Six Thinking Dimensions. At 3.38, the Default outcome indicates a moderate level of performance, but it lacks consistency, depth, and a balanced perspective. It reflects a "middle ground" approach where strengths are diluted and weaknesses remain unaddressed.

## Integrated Result—3.93

The Integrated score, at 3.93, demonstrates a significant improvement over the Default baseline. This result benefits from the orchestrated application of all six dimensions, ensuring evidence is balanced with risk analysis, value recognition, creativity, emotional awareness, and process control. The integrated approach not only raises the overall level of performance but also provides coherence and robustness in decision-making.

## Comparative Analysis

The difference of **+0.55 points** between the Integrated and Default results demonstrates the tangible value of applying the Six Thinking Dimensions systematically. While the Default approach offers a starting reference, it underutilizes available perspectives and leads to uneven outcomes. The Integrated result, by contrast, shows that structured orchestration of the dimensions delivers higher consistency, greater reliability, and a more balanced outcome.

## Final Verdict

The evaluation confirms that the **Integrated method provides better results than the Default**. It transforms fragmented reasoning into a coherent framework, elevating both the quality and robustness of analysis. In practice, this means that organizations or research teams using the Integrated approach can expect stronger decision-making outcomes compared to relying on Default, unstructured reasoning.

The results are derived from a limited set of questions and, therefore, may not fully represent the quality criteria in real-world use cases across all contexts. While the Blue Dimension appears to deliver the strongest results, this is likely influenced by the restricted scope of the evaluation. To ensure applicability across a broader range of problem domains, it remains essential to consider all dimensions collectively rather than privileging one.

The observed weaknesses in the Red (emotion) and Green (creativity) dimensions stem from the inherent limitations of large language models (LLMs). These models do not possess accurate emotional intelligence or creative capacity but instead simulate such behaviors through probabilistic pattern recognition and descriptive knowledge drawn from examples in their training data.

## Closing Thoughts

This research has demonstrated that the **Six Thinking Dimensions** provide a structured and rigorous methodology for expanding evaluative reasoning. Each dimension offers a distinct analytical lens: the **White Dimension** establishes an evidence-based foundation, the **Red Dimension** integrates human emotion and perception, the **Black Dimension** scrutinizes risk and limitations, the **Yellow Dimension** highlights value and potential, the **Green Dimension** introduces creativity and innovation, and the **Blue Dimension** ensures organization, balance, and synthesis.

Experimental results across seven questions reinforce these distinctions. The **Blue Dimension** consistently outperformed others (average score 4.21), confirming its strength in orchestration, structure, and coherence. The **White** and **Yellow Dimensions** both demonstrated high effectiveness (3.91 each), highlighting firm factual grounding and the identification of opportunities. In contrast, the **Red Dimension** (3.01) and the **Green Dimension** (3.11) scored lowest, reflecting the inherent difficulty large language models face in simulating authentic emotional intelligence and divergent creativity. These weaknesses align with the probabilistic and descriptive nature of LLM outputs, which lack actual experiential or affective grounding.

When applied independently, the dimensions generate valuable but partial perspectives. The comparative evaluations confirm that the **Integrated score** (3.93) surpasses the **Default score** (3.38), evidencing that the collective orchestration of all six dimensions provides superior results. This demonstrates that the methodology mitigates blind spots, strengthens consistency, and balances optimism with caution, evidence with emotion, and creativity with structure.

The overall effectiveness of this framework lies in its methodological integration. By systematically incorporating multiple reasoning modalities, the Six Thinking Dimensions transform fragmented viewpoints into a coherent, multilayered analysis. This integrated approach enhances clarity, reveals hidden dependencies, and supports decision-making processes with greater depth, balance, and accountability.

In conclusion, the findings validate that the Six Thinking Dimensions deliver superior evaluative outcomes compared to Default reasoning. Their strength derives not only from the individual contributions of each dimension but also from their collective capacity to function as a holistic system. This positions the framework as both a validated academic contribution and a practical tool for complex domains such as AI-powered agentic ecosystems, where rigorous, transparent, and balanced decision-making is essential.

> **Key Insight**
>
> *The Six Thinking Dimensions form the backbone of monetizing both light and Dark Data for any business.*

## The Future of Six Thinking Dimensions in the Age of AI-Powered Agents

The exponential rise of AI-powered agents with agentic capabilities is reshaping business, society, and even the nature of human cognition. In this evolving landscape, the Six Thinking Dimensions framework does not become obsolete; instead, it undergoes a profound reinvention, extending its value as a meta-cognitive methodology for orchestrating the interplay of human and machine intelligence.

This chapter explores four forward-looking questions:

1. What becomes of the Six Thinking Dimensions as AI systems disrupt established structures?
2. How might the Six Thinking Dimensions evolve between 2025 and 2035?
3. In what ways would artificial general intelligence (AGI) influence their application?
4. Which human skills remain indispensable for maximizing their potential in an AI-augmented world?

## The Six Thinking Dimensions in an Evolving AI Landscape

As AI-powered agents become increasingly agentic, capable of autonomous decision-making, self-adaptation, and swarm-based collaboration, business and social structures experience continuous disruption. Hierarchies flatten, workflows decentralize, and decision cycles accelerate beyond traditional human bandwidth. In this context, the Six Thinking Dimensions serve as an essential orchestration framework:

- **White Dimension (facts and data)** is amplified by AI systems capable of ingesting and analyzing vast data streams in real time.
- **Red Dimension (feelings and intuition)** gains new meaning when AI sentiment analysis and affective computing highlight human emotions at scale.
- **Black Dimension (risks and cautions)** is fortified by predictive models identifying vulnerabilities, biases, and failure modes.
- **Yellow Dimension (benefits and value)** is extended by optimization engines that model positive outcomes and future value generation.
- **Green Dimension (creativity and alternatives)** is accelerated through generative AI systems, producing novel designs, strategies, and solutions.
- **Blue Dimension (process and orchestration)** becomes the control plane for hybrid human-AI orchestration, ensuring balance, governance, and purposeful direction.

Rather than replacing the Six Thinking Dimensions, AI agents will embed them into every layer of decision-making. Each Thinking Dimension becomes an intelligent agentic function, working in harmony with human oversight.

## Predicting the Evolution: 2025–2035

Between 2025 and 2035, the Six Thinking Dimensions will evolve from a human-centered facilitation method into a sociotechnical infrastructure:

1. **2025–2027: Hybrid Augmentation:** Enterprises begin embedding AI-powered Dimensions into decision support platforms, where each Dimension functions as a specialized agent augmenting human teams.
2. **2028–2030: Autonomous Orchestration:** Dimensions operate as autonomous polymorphic agents, dynamically rebalancing decision-making contexts in response to external disruptions (economic, environmental, or geopolitical).
3. **2031–2035: Cognitive Symbiosis:** Humans and AI dimensions achieve coevolutionary cooperation, where the Six Dimensions act as a cognitive compass, guiding not only business decisions but also societal governance, ethics, and cultural adaptation.

By 2035, the Six Thinking Dimensions may represent a universal protocol for structuring cognition in hybrid intelligence ecosystems.

## Implications of Artificial General Intelligence (AGI)

If AGI emerges machines with human-level cognitive flexibility, the Six Thinking Dimensions will become even more critical as a governance and alignment mechanism:

- **Checks and Balances:** AGI systems, like humans, will require structured reasoning modes to prevent bias, overconfidence, or narrow optimization.
- **Ethical Mediation:** The Red and Black Dimensions will be indispensable in encoding moral, cultural, and existential concerns into AGI-driven decisions.
- **Value Alignment:** The Yellow and Blue Dimensions will help align AGI cognition with human-defined goals, sustainability, and fairness.
- **Creative Expansion:** Green Dimension functions will integrate AGI's capacity for generating new paradigms, ensuring innovation remains socially constructive.

Thus, the Six Thinking Dimensions evolve from being a human heuristic to a universal framework for governing distributed cognition across human and machine intelligence.

## Essential Human Skills in the Age of AI Hats

Even as AI augments each Hat, the human-in-the-loop remains indispensable. Four skill domains must be honed:

1. **Interpretive Literacy:** The ability to interpret outputs from AI dimensions critically, distinguishing insight from noise or hallucination.
2. **Ethical Reasoning:** Applying moral frameworks to guide Red and Black Dimension considerations, ensuring alignment with human values.
3. **Orchestration Competence:** Acting as the ultimate Blue Dimension conductor, ensuring that dimensions (both human and AI) are balanced, sequenced, and contextually relevant.
4. **Creative Synthesis:** Leveraging Green and Yellow Dimensions to integrate divergent AI outputs into coherent, innovative strategies that humans can implement.

These human skills transform AI-powered dimensions from a set of tools into a living methodology that scales across enterprises, societies, and governance systems.

## Highlighted Future Statement

> **Future Outlook**
>
> **By 2035, the Six Thinking Dimensions is anticipated to serve as the universal protocol for hybrid human-AI decision-making, uniting artificial and human cognition into a coherent, ethical, and innovation-driven future. It envisions a world where collective human intelligence and collective machine intelligence collaborate synergistically, enhancing one another's strengths while safeguarding against each other's inherent limitations.**

> **Warning Outlook**
>
> **If humans fail to orchestrate AI-powered dimensions effectively, decision-making may fragment into bias-amplifying silos. Without strong human-in-the-loop oversight, hybrid cognition could accelerate unethical outcomes, systemic instability, and loss of human agency.**

## Closing Perspective

The Six Thinking Dimensions, far from finalized, are entering their most transformative phase. From 2025 to 2035, they will shift from facilitation heuristics into the backbone of hybrid intelligence governance. In the era of AGI, they may form a part of humanity's most reliable scaffolding for aligning machine cognition with human flourishing.

Looking ahead, the Six Dimensions should evolve into a universal language for structured reasoning, not only guiding business strategy but also shaping cultural, societal, and even planetary decision-making. Its role may expand from a boardroom methodology to a global framework for effective orchestration of hybrid intelligence.

The task ahead is not to abandon the dimensions but to remaster them, embracing their evolution as intelligent companions and governance anchors in the AI age. By doing so, we ensure that humans remain the conductors of a future symphony where human creativity and machine intelligence blend into harmony, unlocking unprecedented innovation, resilience, and prosperity for generations to come.

The Six Thinking Dimensions could be both *a compass and a map*: the compass that keeps humanity oriented toward purposeful goals and the map that charts pathways through the complexity of hybrid cognition. With them, we can navigate the unknown not with fear, but with clarity, imagination, and hope.

> **Call to Action**
>
> **The future demands clarity, courage, and orchestration. Now is the time to reengage with the Six Thinking Dimensions serving as our guiding framework to manage disruption, unearth creativity, and craft a future where both humanity and AI flourish in harmony.**

> **Unified Perspective**
>
> **Together, these appendixes transform theory into practice, linking structured evaluation, data transformation, and business monetization into a single continuum of applied intelligence.**

## Closing Words

As you reach the final page of this book, I want to thank you for investing your time, energy, and curiosity in exploring the frontier of AI-powered agents and the Six Thinking Dimensions. The journey you have taken through these chapters has not only been an exploration of frameworks and theories, but also a call to reimagine how humans and machines can think, decide, and create together.

The field you now stand in is one of the most dynamic, disruptive, and transformative domains in human history. From decision orchestration to agentic intelligence, you have gained insights that are not merely academic but efficient tools for shaping the future of business, society, and governance.

But knowledge alone is never enough. The actual test begins with what you choose to do next. I dare you to take the concepts, methods, and visions presented here and apply them boldly, experiment with them, challenge them, and extend them. Build systems that matter. Lead teams that harness both human and machine intelligence. Create frameworks that will guide not only your organization, but also the next generation of thinkers, leaders, and innovators.

> **Call to Action**
>
> **The responsibility and the opportunity now rest with you.**

> **Good Wishes**
>
> **May the Six Thinking Dimensions be both your compass and your map as you journey through the uncharted horizons of AI-powered agents. With courage to explore, creativity to build, and orchestration to unite, you hold the power not merely to take part in this transformation, but to lead it, shaping a future where human imagination and machine intelligence flourish together.**

> **Imagination is more important than knowledge.**
> **For knowledge is limited,**
> **whereas imagination embraces the entire world,**
> **stimulating progress,**
> **giving birth to evolution.**
>
> — Albert Einstein

# Appendixes

## Appendixes Overview

The appendixes provide supporting material that extends the theoretical and practical discussions presented in the main chapters. Each appendix is designed to deepen understanding, provide structured evaluations, or illustrate applied frameworks for data-driven decision-making. Together, they form the evidence base and practical toolkit accompanying the Six Thinking Dimensions in action.

## *Appendixes 1–7: Evaluation of Results (Linked to Chap. 1)*

Appendixes 1 to 7 contain the detailed evaluation results generated from the experiments presented in Chap. 1. Each appendix documents the structured scoring of responses across the Six Thinking Dimensions framework, using quantitative and qualitative metrics. These evaluations provide transparency into the decision-analysis process, allowing readers to trace how evidence, risks, opportunities, creativity, and orchestration were measured and synthesized into conclusions. They form the empirical foundation upon which the chapter's interpretations were constructed.

## *Appendix 8: From Dark Data to Light Data to Monitored Data*

Appendix 8 outlines a transformation framework for organizational data maturity. It describes how enterprises can identify hidden or unused *Dark Data*, convert it into structured and actionable *Light Data*, and embed it into continuous monitoring pipelines to create *monitored data*. The appendix presents guiding questions,

© The Author(s), under exclusive license to APress Media, LLC,      125
part of Springer Nature 2026
A. F. Vermeulen, *Agentic Hyper-Personalized Dimensions*,
https://doi.org/10.1007/979-8-8688-1877-6

pathways, and practical examples for uncovering untapped insights. By doing so, it demonstrates how often neglected data can be systematically monetized, enhancing efficiency, compliance, and innovation capacity.

## Appendix 9: Building a Profit Playbook for SMEs

Appendix 9 introduces a structured playbook for small- and medium-sized enterprises (SMEs) to monetize both light and Dark Data. It reframes every business department, such as sales, marketing, finance, operations, IT, and HR, not as cost centers, but as potential contributors to profitability. Through a series of guiding questions and frameworks, the appendix shows how each function can connect decision-making to measurable profit outcomes. In particular, it demonstrates how SMEs can integrate data insights into daily operations to create a holistic profit-maximization strategy. This playbook serves as a practical roadmap for leaders aiming to turn overlooked data into tangible business value.

## Appendix 10: Business Cases—Industry Sectors

This appendix presents a set of carefully selected business cases drawn from diverse industry sectors, including AIOps, healthcare, finance, public sector, citizen services, and space exploration, to illustrate how the Six Thinking Dimensions framework can be applied to unlock value from business Dark Data. For each sector, we dissect the core business challenges, demonstrate how each thinking dimension contributes to intelligent, hyper-personalized agentic systems, and outline the potential monetization opportunities. Together, these cases offer a rich, multidimensional view of how Agentic AI can transform real-world enterprise contexts.

## Appendix 11: Business Cases—UN Sustainable Development Goals

This appendix extends the Six Thinking Dimensions framework into the realm of global development by exploring 17 business use cases, one corresponding to each of the United Nations Sustainable Development Goals (SDGs). Each case examines a pressing population-driven societal challenge, details how the Six Thinking Dimensions can meaningfully contribute to multidimensional solutions, and highlights the economic, social, or strategic benefits that may arise. By combining enterprise-grade AI with population and sustainability imperatives, these cases demonstrate how technology and global impact can converge for systemic, scalable transformation.

# Appendix 1

## Question

In the context of modern healthcare, should AI systems be allowed to make medical diagnoses alongside human doctors? Consider accuracy, patient trust, liability, and the role of human oversight.

## Results

### *Evaluation of Responses—Using the Six Thinking Dimensions*

The **Evaluation of Responses for Using the Six Thinking Dimensions** applies **Table A-1 — Scoring Rubric for Evaluating Responses Across the Six Thinking Dimensions**, which outlines the assessment structure for analysing responses across all six dimensions.

**Scoring Rubric**

**Scale:** 1=poor, 2=fair, 3=good, 4=very Good, 5=excellent.

| Criterion | Description |
| --- | --- |
| Persona fidelity (PF) | Adherence to the specified role's content and style constraints. |
| Coverage (COV) | Completeness relative to the remit (e.g., accuracy, trust, liability, oversight). |
| Specificity (SPEC) | Use of precise, concrete details; avoidance of generic statements. |
| Actionability (ACT) | Presence of clear, feasible next steps where appropriate. |
| Distinctiveness (DIST) | Clear differentiation from other personas' outputs. |
| Clarity (CLAR) | Organization, readability, and length discipline. |

**Table A-1** Scoring Rubric Used for Evaluating Responses Across the Six Thinking Dimensions

## *White Dimension—Scorecard and Appraisal*

| Criterion | Score (1–5) |
| --- | --- |
| PF | 4 |
| COV | 5 |
| SPEC | 4 |
| ACT | 4 |
| DIST | 5 |
| CLAR | 5 |
| **Overall** | **4.50** |

**Table A-2** Scorecard for the White Dimension

**Justification:** Presents an evidence-led analysis of diagnostic accuracy, error modes, data quality, and validation requirements with phased implementation and auditability; emphasizes human-in-the-loop control. **Strengths:** Clear structure; balanced treatment of accuracy, trust, liability, and oversight; explicit recommendations. **Weaknesses:** Lacks quantitative benchmarks (e.g., sensitivity/specificity vs. standard-of-care) and explicit KPIs.

## *Red Dimension—Scorecard and Appraisal*

The **Red Dimension — Scorecard and Appraisal** is evaluated using **Table A-3 — Scorecard for the Red Dimension.**

| Criterion | Score (1–5) |
| --- | --- |
| PF | 5 |
| COV | 2 |
| SPEC | 2 |
| ACT | 2 |
| DIST | 5 |
| CLAR | 4 |
| **Overall** | **3.33** |

**Table A-3** Scorecard for the Red Dimension

**Justification:** Conveys stakeholder emotions (hope, anxiety, perceived fairness) and insists on transparency and bedside communication but offers limited concrete steps. **Strengths:** Authentic emotive register; highlights trust and consent concerns. **Weaknesses:** Sparse criteria for acceptance; lacks patient-engagement protocols and measurement of sentiment.

## *Black Dimension—Scorecard and Appraisal*

The **Black Dimension — Scorecard and Appraisal** is evaluated using **Table A-4 — Scorecard for the Black Dimension**.

| Criterion | Score (1–5) |
| --- | --- |
| PF | 5 |
| COV | 5 |
| SPEC | 4 |
| ACT | 4 |
| DIST | 5 |
| CLAR | 5 |
| **Overall** | **4.67** |

**Table A-4** Scorecard for the Black Dimension

**Justification:** Foregrounds risks (bias, misdiagnosis, distribution shift, automation complacency, liability exposure) and proposes mitigations (tiered use, validation, transparency, audits, pilot gating). **Strengths:** Thorough risk taxonomy; structured mitigations; strong oversight framing. **Weaknesses:** Few quantitative stop/go thresholds; legal doctrines cited only generally.

## *Yellow Dimension—Scorecard and Appraisal*

| Criterion | Score (1–5) |
|-----------|-------------|
| PF | 5 |
| COV | 5 |
| SPEC | 3 |
| ACT | 4 |
| DIST | 5 |
| CLAR | 5 |
| **Overall** | **4.50** |

**Table A-5** Scorecard for the Yellow Dimension

**Justification:** Highlights benefits (earlier detection, workload reduction, consistency, personalization) with a pragmatic adoption path under human oversight. **Strengths:** Clear value narrative; phased rollout; attention to clinician enablement. **Weaknesses:** Limited quantified benefit models (e.g., time-to-diagnosis, avoided adverse events).

## *Green Dimension—Scorecard and Appraisal*

The **Green Dimension — Scorecard and Appraisal** is evaluated using **Table A-6 — Scorecard for the Green Dimension**.

| Criterion | Score (1–5) |
|-----------|-------------|
| PF | 5 |
| COV | 4 |
| SPEC | 3 |
| ACT | 3 |
| DIST | 5 |
| CLAR | 4 |
| **Overall** | **4.00** |

**Table A-6** Scorecard for the Green Dimension

**Justification:** Reframes the problem toward *augmented* rather than *autonomous* diagnosis; proposes creative approaches (explainability visuals, differential-diagnosis copilots, simulation-based training), with moderate operational detail. **Strengths:** Conceptual provocation; keeps human factors central. **Weaknesses:** Requires more precise mechanisms, pilots, and measurement plans.

## *Blue Dimension—Scorecard and Appraisal*

The **Blue Dimension — Scorecard and Appraisal** is evaluated using **Table A-7 — Scorecard for the Blue Dimension**.

| Criterion | Score (1–5) |
|-----------|-------------|
| PF | 5 |
| COV | 4 |
| SPEC | 3 |
| ACT | 4 |
| DIST | 5 |
| CLAR | 5 |
| **Overall** | **4.33** |

**Table A-7** Scorecard for the Blue Dimension

**Justification:** Excels at orchestration: scope definition, governance cadence, phased deployment, monitoring, and "next steps" aligned to process control and balance; specificity could be deeper. **Strengths:** Clear facilitation; iterative learning loops; governance alignment. **Weaknesses:** Lacks concrete artefacts (RACI, RAID, cadence calendar) and fully specified metrics.

## *Comparative Summary Across Dimensions*

The **Comparative Summary of Evaluation Results** across all Six Thinking Dimensions is presented in **Table A-8 — Comparative Summary of Evaluation Results Across the Six Thinking Dimensions**.

| Dimension | PF | COV | SPEC | ACT | DIST | CLAR | Overall |
|-----------|-----|-----|------|-----|------|------|---------|
| White | 4 | 5 | 4 | 4 | 5 | 5 | 4.50 |
| Red | 5 | 2 | 2 | 2 | 5 | 4 | 3.33 |
| Black | 5 | 5 | 4 | 4 | 5 | 5 | 4.67 |
| Yellow | 5 | 5 | 3 | 4 | 5 | 5 | 4.50 |
| Green | 5 | 4 | 3 | 3 | 5 | 4 | 4.00 |
| Blue | 5 | 4 | 3 | 4 | 5 | 5 | 4.33 |

**Table A-8** Comparative Summary of Evaluation Results Across the Six Thinking Dimensions

**Synthesis:** *Top performers* by overall score are **Black** (4.67), **White** (4.50), and **Yellow** (4.50). Together, they provide a robust backbone for AI-assisted diagnosis: risk containment and liability awareness (Black), evidence discipline and validation (White), and value framing with clinician workload relief and earlier detection (Yellow). **Blue** (4.33) offers strong orchestration and should add concrete governance artefacts. **Green** (4.00) supplies creative augmentation concepts needing sharper pilots and metrics. **Red** (3.33) effectively surfaces stakeholder emotions and trust dynamics but needs actionable acceptance criteria.

## *Next-Step Adoption Guide for AI-Assisted Diagnosis*

- **Define KPIs and Thresholds:** Sensitivity, specificity, PPV/NPV by indication; noninferiority margins vs. standard-of-care; calibration and drift metrics
- **Human Oversight by Design:** Mandate clinician verification for high-risk decisions; require explainability artefacts at the point of care
- **Governance and Liability:** Establish RACI across developers, providers, and clinicians; maintain audit trails; align indemnity and incident response
- **Deployment Phasing:** Sandbox → shadow mode → assist mode → tightly scoped codiagnosis; gate progress on KPI attainment and patient safety reviews
- **Trust and Communication:** Informed consent language, model cards, and patient-facing explanations; measure sentiment and comprehension
- **Continuous Validation:** Real-world performance monitoring, bias audits by subgroup, and periodic recertification of models and data pipelines

## *Summary and Advisory on Evaluation Results*

The combined analysis supports *allowing AI to participate alongside human doctors* under a governed, phased, and measurable framework. Evidence-first structuring (White) and rigorous risk controls (Black) are prerequisites to protect patients and manage liability. Opportunity framing (Yellow) shows clear potential in earlier detection, consistency, and clinician workload relief when human oversight is mandated. Creative augmentation (Green) should focus on explainability and copilot workflows rather than autonomy. Process orchestration (Blue) must codify roles, gates, cadence, and KPI stacks to sustain safe adoption. Finally, trust (Red) requires intentional communication, consent, and ongoing sentiment measurement. Under these conditions, AI-assisted diagnosis can enhance care quality while preserving patient trust and clinician accountability.

# Appendix 2

## Question

Should governments impose a carbon tax on all businesses as a strategy to combat climate change? Discuss economic impact, fairness, risks of unintended consequences, and long-term benefits.

## Results

### *Evaluation Overview*

This research-grade evaluation reviews six persona-specific responses (White, Red, Black, Yellow, Green, Blue) to Question 2 on economy-wide carbon taxation. Each response is scored against a standard rubric and accompanied by qualitative analysis, highlighting strengths, weaknesses, and concrete improvements.



A. F. Vermeulen, *Agentic Hyper-Personalized Dimensions*,
https://doi.org/10.1007/979-8-8688-1877-6

## *Scoring Rubric*

The **Scoring Rubric for Evaluating Persona Responses** is presented in
Table A-9 — Scoring Rubric Criteria for Evaluating Persona Responses.
**Scale:** 1=poor, 2=fair, 3=good, 4=very good, 5=excellent.

| | |
|---|---|
| **Persona fidelity (PF)** | Adherence to the specified role's content and style constraints. |
| **Coverage (COV)** | Completeness relative to the persona's remit (e.g., risks, facts, opportunities). |
| **Specificity (SPEC)** | Use of precise, concrete details; avoidance of generic statements. |
| **Actionability (ACT)** | Presence of clear, feasible next steps where appropriate. |
| **Distinctiveness (DIST)** | Clear differentiation from other personas' outputs. |
| **Clarity (CLAR)** | Organization, readability, and length discipline. |

**Table A-9** Scoring Rubric Criteria for Evaluating Persona Responses

## *White Dimension—Scorecard and Analysis*

The **White Dimension — Scorecard and Appraisal** is evaluated using
Table A-10 — White Dimension Scorecard for Question 2.

| Criterion | PF | COV | SPEC | ACT | DIST | CLAR | Overall |
|---|---|---|---|---|---|---|---|
| Score (1–5) | 3 | 4 | 2 | 2 | 3 | 4 | 3.00 |

**Table A-10** White Dimension Scorecard for Question 2

**Justification and Evidence:** The response offers a balanced overview across economic impact, fairness, risks (e.g., leakage, inflation), and long-term benefits and references examples like Sweden, Canada/British Columbia, and the IPCC in general terms. However, it relies heavily on uncited claims, contains overstatements (e.g., temperature "decline" rather than reduced warming relative to baseline), and lacks quantitative anchors (no tax-rate ranges, revenue magnitudes, distributional numbers). Persona fidelity is partial: it aims at factual balance but falls short on sourcing and precision expected of White.

## *Red Dimension—Scorecard and Analysis*

The **Red Dimension — Scorecard and Appraisal** is evaluated using
Table A-11 — Red Dimension Scorecard for Question 2.

| Criterion | PF | COV | SPEC | ACT | DIST | CLAR | Overall |
|-----------|-----|-----|------|-----|------|------|---------|
| Score (1–5) | 5 | 3 | 1 | 1 | 5 | 3 | 3.00 |

**Table A-11**  Red Dimension Scorecard for Question 2

## *Black Dimension—Scorecard and Analysis*

The **Black Dimension — Scorecard and Appraisal** is evaluated using **Table A-12 — Black Dimension Scorecard for Question 2**.

| Criterion | PF | COV | SPEC | ACT | DIST | CLAR | Overall |
|-----------|-----|-----|------|-----|------|------|---------|
| Score (1–5) | 5 | 4 | 2 | 4 | 4 | 4 | 3.83 |

**Table A-12**  Black Dimension Scorecard for Question 2

## *Yellow Dimension—Scorecard and Analysis*

The **Yellow Dimension — Scorecard and Appraisal** is evaluated using **Table A-13 — Yellow Dimension Scorecard for Question 2**.

| Criterion | PF | COV | SPEC | ACT | DIST | CLAR | Overall |
|-----------|-----|-----|------|-----|------|------|---------|
| Score (1–5) | 5 | 4 | 2 | 4 | 4 | 4 | 3.83 |

**Table A-13**  Yellow Dimension Scorecard for Question 2

## *Green Dimension—Scorecard and Analysis*

The **Green Dimension — Scorecard and Appraisal** is evaluated using **Table A-14 — Green Dimension Scorecard for Question 2**.

| Criterion | PF | COV | SPEC | ACT | DIST | CLAR | Overall |
|-----------|-----|-----|------|-----|------|------|---------|
| Score (1–5) | 5 | 4 | 2 | 3 | 5 | 4 | 3.83 |

**Table A-14**  Green Dimension Scorecard for Question 2

## *Blue Dimension—Scorecard and Analysis*

The **Blue Dimension — Scorecard and Appraisal** is evaluated using **Table A-15 — Blue Dimension Scorecard for Question 2**.

| Criterion | PF | COV | SPEC | ACT | DIST | CLAR | Overall |
|---|---|---|---|---|---|---|---|
| Score (1–5) | 5 | 4 | 2 | 4 | 4 | 4 | 3.83 |

**Table A-15** Blue Dimension Scorecard for Question 2

## *Comparative Summary Across Dimensions*

**Overall Pattern:** Black, Yellow, Green, and Blue cluster at ≈ 3.83 overall, each excelling in its role yet constrained by low specificity. White and Red average 3.00 for different reasons: White is broad but underevidenced; Red is affectively strong but nonactionable.

## *Summary and Advisory*

The evaluation of Question 2 through the Six Thinking Dimensions revealed a broad spectrum of strengths and weaknesses. The White Dimension demonstrated an adequate balance of facts, risks, and benefits but suffered from insufficient citations and a lack of quantitative specificity. While it provided a structured overview, the absence of hard evidence reduced persona fidelity and left its analysis vulnerable to overgeneralization. The Red dimension successfully captured emotional and ethical sensitivities, particularly concerning fairness and the plight of low-income households, but did so at the expense of actionable guidance and data-driven clarity.

The Black and Yellow Dimensions provided the most policy-relevant material, both achieving strong alignment with their respective mandates. Black delivered a disciplined risk framework, identifying potential pitfalls such as regressivity, political backlash, and carbon leakage, while also recommending phased rollouts and monitoring mechanisms. Yellow contributed a value-oriented perspective, emphasizing innovation, public health, and long-term gains. However, both dimensions lacked quantitative thresholds and sector-specific data, which are essential for turning strategic insights into implementable policies.

The Green Dimension stood out for its creativity, introducing alternative mechanisms such as carbon credit currencies and circular economy incentives. Its imaginative approach provided useful provocations for future innovation in climate policy. However, feasibility analysis and governance structures were limited, reducing their practical utility in immediate policymaking. By contrast, the Blue Dimension offered strong process orchestration, highlighting sequencing, monitoring, and facilitation. Despite its effective balance and structure, it too lacked governance cadence and measurable decision gates.

Taken together, the evaluations show that while each dimension achieved moderate to high overall performance, the most persistent shortfall across all personas was the lack of specificity and quantified detail. Assertions were often presented without numerical backing, thresholds, or reference to empirical case studies. This hinders the integration of these perspectives into a cohesive, evidence-based carbon tax strategy. Furthermore, while Red captured the human dimension effectively, its lack of actionability suggests it should be paired with Black's risk awareness and Yellow's optimism for maximum policy coherence.

The advisory outcome is clear: to create a robust, fair, and effective carbon tax framework, future outputs should integrate White's factual scaffolding with Black's safeguards, Yellow's opportunities, Green's creativity, and Blue's orchestration. This synthesis requires the systematic addition of quantified metrics (price paths, revenue models, CPI guardrails) and independent governance mechanisms. With these enhancements, the collective output of the Six Dimensions can evolve into a balanced, adaptive, and socially legitimate blueprint for carbon taxation policy.

# Appendix 3

## Question

Should a global company transition permanently to a remote-first workplace model? Explore implications for productivity, culture, innovation, employee well-being, and international competitiveness.

## Results

### *Evaluation of Responses for Question 3*

**Question:** *Should a global company transition permanently to a remote-first workplace model? Explore implications for productivity, culture, innovation, employee well-being, and global competitiveness. Discuss economic impact, fairness, risks of unintended consequences, and long-term benefits. Consider risks, potential opportunities, ethical implications, and creative approaches.*

### *Scoring Rubric*

The **Scoring Rubric for Evaluating Responses to Question 3** is presented in **Table A-16 — Scoring Rubric Criteria for Evaluating Responses to Question 3**. **Scale:** 1=poor, 2=fair, 3=good, 4=very good, 5=excellent.



A. F. Vermeulen, *Agentic Hyper-Personalized Dimensions*, https://doi.org/10.1007/979-8-8688-1877-6

| Criterion | Description |
|---|---|
| **Persona fidelity (PF)** | Adherence to the specified role's content and style constraints. |
| **Coverage (COV)** | Completeness relative to the persona's remit (e.g., risks, facts, opportunities). |
| **Specificity (SPEC)** | Use of precise, concrete details; avoidance of generic statements. |
| **Actionability (ACT)** | Presence of clear, feasible next steps where appropriate. |
| **Distinctiveness (DIST)** | Clear differentiation from other personas' outputs. |
| **Clarity (CLAR)** | Organization, readability, and length discipline. |

**Table A-16** Scoring Rubric Criteria for Evaluating Responses to Question 3

## *White Dimension Evaluation*

The **White Dimension — Scorecard and Appraisal** is evaluated using **Table A-17 — White Dimension Evaluation Scores for Question 3**.

| Criteria | PF | COV | SPEC | ACT | DIST | CLAR |
|---|---|---|---|---|---|---|
| **Score** | 5 | 5 | 5 | 4 | 5 | 5 |

**Table A-17** White Dimension Evaluation Scores for Question 3

**Justification:** Strong reliance on evidence, citing credible studies, quantifying uncertainties, and highlighting verified data. Specificity was high with concrete metrics (e.g., 5–15% productivity impacts in different contexts). Actionability was present but leaned toward strategic recommendations rather than immediate steps. Excellent persona fidelity and clarity. Improvement could include more structured comparative tables.

## *Red Dimension Evaluation*

The **Red Dimension — Scorecard and Appraisal** is evaluated using **Table A-18 — Red Dimension Evaluation Scores for Question 3**.

| Criteria | PF | COV | SPEC | ACT | DIST | CLAR |
|---|---|---|---|---|---|---|
| **Score** | 5 | 3 | 3 | 2 | 5 | 4 |

**Table A-18** Red Dimension Evaluation Scores for Question 3

**Justification:** Excellent emotional resonance, intuition, and instinctual reactions, reflecting persona fidelity. Coverage of domains was uneven, with a strong focus on loneliness and culture but lacking balance. Specificity was limited, and actionability was weak (warnings rather than steps). Distinctiveness was strong due to its emotive framing. Suggested improvement: integrate concrete emotional metrics (surveys, sentiment analysis).

## *Black Dimension Evaluation*

The **Black Dimension — Scorecard and Appraisal** is evaluated using **Table A-19 — Black Dimension Evaluation Scores for Question 3**.

| Criteria | PF | COV | SPEC | ACT | DIST | CLAR |
|---|---|---|---|---|---|---|
| Score | 5 | 5 | 4 | 5 | 5 | 5 |

**Table A-19**  Black Dimension Evaluation Scores for Question 3

**Justification:** Robust risk assessment with probability/severity levels and detailed mitigations. Coverage was comprehensive across productivity, culture, innovation, well-being, and competitiveness. Actionability was high with phased recommendations. Specificity was strong though occasionally generalized (e.g., "robust monitoring systems"). Distinctiveness was clear in its cautionary stance. Minor improvement: quantify risks more consistently.

## *Yellow Dimension Evaluation*

The **Yellow Dimension — Scorecard and Appraisal** is evaluated using **Table A-20 — Yellow Dimension Evaluation Scores for Question 3**.

| Criteria | PF | COV | SPEC | ACT | DIST | CLAR |
|---|---|---|---|---|---|---|
| Score | 5 | 5 | 4 | 4 | 5 | 5 |

**Table A-20**  Yellow Dimension Evaluation Scores for Question 3

**Justification:** Highlighted opportunities in talent acquisition, productivity, innovation, and well-being. Coverage was excellent with future-oriented optimism. Specificity was reasonable but less evidence-heavy than White. Actionability is present via phased pilots, tech infrastructure, and task force creation. Distinctive optimism maintained persona fidelity. Suggested improvement: link opportunities to quantified metrics.

## *Green Dimension Evaluation*

The **Green Dimension — Scorecard and Appraisal** is evaluated using **Table A-21 — Green Dimension Evaluation Scores for Question 3**.

| Criteria | PF | COV | SPEC | ACT | DIST | CLAR |
|----------|----|-----|------|-----|------|------|
| Score | 5 | 4 | 4 | 4 | 5 | 4 |

**Table A-21**  Green Dimension Evaluation Scores for Question 3

**Justification:** Excelled in creativity and provocation (e.g., "digital silence," "analog revival"). Persona fidelity was excellent with radical alternatives. Coverage was strong but less systematic than White/Black. Actionability was practical (pilots, feedback loops) but framed as prototypes. Clarity was slightly reduced by playful tone. Suggested improvement: balance creative provocation with structured comparative analysis.

## *Blue Dimension Evaluation*

The **Blue Dimension — Scorecard and Appraisal** is evaluated using **Table A-22 — Blue Dimension Evaluation Scores for Question 3**.

| Criteria | PF | COV | SPEC | ACT | DIST | CLAR |
|----------|----|-----|------|-----|------|------|
| Score | 5 | 5 | 4 | 5 | 5 | 5 |

**Table A-22**  Blue Dimension Evaluation Scores for Question 3

**Justification:** Provided metacognitive facilitation, process orientation, and orchestration. Persona fidelity was excellent. Coverage spanned productivity, culture, innovation, well-being, and competitiveness. Specificity was reasonable but oriented to structured facilitation rather than data depth. Actionability was clear (pilots, metrics, governance). Distinctiveness is strong as an orchestrator. Suggested improvement: include comparative framework visuals.

## *Comparative Summary*

Across dimensions, the responses aligned well with their roles. White and Black were the most evidence-based and risk-aware, respectively, scoring highest in specificity and caution. Yellow and Green emphasized opportunities and creativity, with high distinctiveness but slightly less structured evidence. Red captured the emotive human element but lacked actionable depth. Blue provided orchestration and balance, excelling in clarity and process control. Together, they present a holistic, research-grade evaluation of the remote-first transition.

## *Summary and Advisory Insights*

The evaluation of the six dimensions for this question demonstrates strong alignment between persona roles and their intended purposes. The **White Dimension** excelled at evidence-based insight, presenting quantified uncertainties and verified studies, setting a high standard for factual rigor. Objective, data-driven approaches provide a strong foundation for decision-making and would benefit from concise comparative tables.

The **Red Dimension** effectively conveyed the emotional weight of a remote-first transition, giving voice to instinctive reactions often overlooked in strategic debate. To increase practical value, pair these intuitions with sentiment surveys, pulse checks, and well-being indices to translate signals into actions.

The **Black Dimension** offered comprehensive, structured risk analysis. By categorizing risks into probability and severity and pairing each with targeted mitigations, it delivered high actionability and clarity. Future improvement could include scenario simulations to quantify exposure more consistently.

The **Yellow and Green Dimensions** provided complementary strengths: optimism and creativity. Yellow highlighted opportunities in productivity, talent, and innovation with phased steps; Green provoked radical rethinking through playful experiments. Both perspectives are vital for innovation, and both benefit from embedding empirical validation to balance aspiration with evidence.

Finally, the **Blue Dimension** acted as the orchestral facilitator, synthesizing viewpoints into a process-oriented roadmap. Recommendations for pilots, metrics, and governance underscore the value of structured balance when navigating transformation. Used together, the six dimensions—evidence (White), emotion (Red), caution (Black), optimism (Yellow), creativity (Green), and orchestration (Blue)—enable a phased, evidence-based, and human-centered approach to adopting a remote-first model.

# Appendix 4

## Question

How should a large retailer ethically monetize customer behavior data? Address privacy regulations, trust, risks, opportunities, and potential innovative approaches to value creation.

## Results

### *Scoring Rubric*

The **Scoring Rubric for Evaluating Responses to Q04** is presented in **Table A-23 — Scoring Rubric for Q04 Evaluation**.
**Scale:** 1=poor, 2=fair, 3=good, 4=very good, 5=excellent.

| Criterion | Description |
| --- | --- |
| **Persona fidelity (PF)** | Adherence to the specified role's content and style constraints. |
| **Coverage (COV)** | Completeness relative to the persona's remit (e.g., risks, facts, opportunities). |
| **Specificity (SPEC)** | Use of precise, concrete details; avoidance of generic statements. |
| **Actionability (ACT)** | Presence of clear, feasible next steps where appropriate. |
| **Distinctiveness (DIST)** | Clear differentiation from other personas' outputs. |
| **Clarity (CLAR)** | Organization, readability, and length discipline. |

Table A-23 Scoring Rubric for Q04 Evaluation

## *Quantitative Scores for Q04*

The **Quantitative Evaluation Scores for Q04** across all Six Thinking Dimensions are presented in **Table A-24 — Scorecard for Q04 Across the Six Thinking Dimensions**.
**Criteria:** PF, COV, SPEC, ACT, DIST, CLAR, Overall (mean).

| Persona | PF | COV | SPEC | ACT | DIST | CLAR | Overall |
|---------|----|-----|------|-----|------|------|---------|
| Default | 3 | 4 | 3 | 3 | 2 | 4 | 3.17 |
| White | 5 | 5 | 4 | 3 | 4 | 5 | 4.33 |
| Red | 5 | 3 | 2 | 2 | 4 | 3 | 3.17 |
| Black | 5 | 4 | 4 | 4 | 4 | 4 | 4.17 |
| Yellow | 4 | 4 | 3 | 3 | 3 | 4 | 3.50 |
| Green | 4 | 4 | 3 | 3 | 4 | 4 | 3.67 |

**Table A-24**  Scorecard for Q04 Across the Six Thinking Dimensions

## *Dimension-by-Dimension Evaluation (Evidence-Based)*

White (4.33, strongest overall)

Excellent legal grounding (GDPR/CCPA/PIPEDA), privacy-by-design emphasis, and a balanced inventory of value levers (personalization, supply chain optimization, federated learning, differential privacy). Actionability lags—convert principles into a concrete rollout plan and KPIs.

Black (4.17)

Clear risk doctrine with phased monetization, explicit prohibitions (no direct sale of identifiable data), concrete controls (independent ethics board, breach playbooks, "sunset clauses"), and technical mitigations (differential privacy, federated learning). Strengthen measurement (risk appetite statements, thresholds) and add red teaming for reidentification risk.

Green (3.67)

Strong innovation pipeline (behavioral insights-as-a-service, predictive loyalty, mood-aware support) tempered with privacy tooling (differential privacy, synthetic data). Needs sharper guardrails and go/no-go criteria for high-risk ideas (e.g., sentiment/mood inference). Tie each idea to a testable business case and an ethics checklist.

Yellow (3.50)

Compelling value narrative (personalized experiences, proactive service, supply chain efficiency) with practical safeguards (bias audits, anti-dark-pattern UX). Improve specificity on consent flows and add quantitative benefit–risk scoring to prioritize opportunities.

Red (3.17)

Authentic articulation of stakeholder feelings (unease, manipulation risk) and trust fragility; proposes empathetic "value exchange" and decentralized control. However, it is light on concrete steps and technical specifics—convert sentiments into measurable trust signals and pilot criteria.

Default (3.17, baseline)

Broad coverage across privacy, security, bias, and a menu of strategies (contextual ads, loyalty, insights, sales, synthetic data). Distinctiveness is low and steps are generic; tighten to retailer context with explicit controls, metrics, and a milestone plan.

## *Cross-Dimension Synthesis*

- **What Is Strong:** Legal/evidentiary rigor (White), operational risk discipline (Black), and a credible innovation funnel (Green) form a solid triad for ethical monetization.
- **What Is Missing:** A unified, testable operating model that turns principles (White), cautions (Black), and ideas (Green/Yellow) into pilots with *explicit* consent UX, privacy-tech baselines, and trust KPIs. Red highlights trust sentiment but lacks instrumentation to observe and react.

## *Actionable Recommendations (Next 30–90 Days)*

1. **Adopt a phased monetization playbook (Black → White).** Phase 1: anonymized, aggregated insights for internal use; Phase 2: opt-in personalization; Phase 3: partner insights under differential privacy. Each phase requires DPIAs, model cards, and an ethics sign-off.
2. **Stand up a privacy engineering baseline.** Mandate differential privacy noise budgets, k-anonymity checks, and federated learning where feasible; maintain an internal "reidentification red team" cadence.
3. **Design consent and control UX (Yellow → White).** Ship a granular consent center (purpose-level toggles, data download/delete), ban dark patterns, and publish a living "data use ledger" customers can view. Measure opt-in rates, revocations, and time-to-fulfil deletion.
4. **Pilot two Green ideas with Red guardrails.** (i) *Predictive loyalty* (low–moderate risk). (ii) *Insights-as-a-service* using synthetic/differentially private data. Precommit to stop/go criteria tied to trust KPIs (complaints, opt-out spikes).
5. **Institutionalize governance.** Constitute a cross-functional Data Ethics Committee with veto power; publish model cards and impact assessments; add bias audits to MLOps gates.

## *Verdict*

The White and Black personas supply a robust, lawful and risk-aware backbone; Green and Yellow add pragmatic innovation and value creation; Red rightly centers human trust but needs operationalization. Converged, they support an *ethical, privacy-first, opt-in* monetization roadmap that is both defensible and value-positive for a large retailer.

# Appendix 5

## Question

Should humanity invest heavily in colonizing Mars within this century? Evaluate feasibility, risks, costs, societal inspiration, and alternative ways of ensuring humanity's long-term survival.

## Results

### *Evaluation Overview (Question 5: Mars Colonization)*

Using the provided rubric (PF, COV, SPEC, ACT, DIST, CLAR; 1–5 scale), I evaluated each uploaded response across the Six Dimensions (plus Default). Overall quality is strong on risk framing (Black) and orchestration (Blue), with optimistic value framing (Yellow) also robust. Green brings novel ideas but needs sharper operationalization. White is balanced but could cite more concrete evidence chains. Red is emotionally faithful yet light on decision-ready specifics. Default is serviceable baselining but undifferentiated.

### *Quantitative Scoring Table*

The **Quantitative Evaluation Scores for Q05 (Mars Colonization)** across all Six Thinking Dimensions are presented in **Table A-25 — Scorecard for Q05 (Mars Colonization) Across the Six Thinking Dimensions**.

**Scale:** 1=poor, 2=fair, 3=good, 4=very good, 5=excellent.

| Persona | PF | COV | SPEC | ACT | DIST | CLAR | Overall |
|---|---|---|---|---|---|---|---|
| Black | 5 | 4 | 4 | 5 | 5 | 5 | 4.67 |
| Blue | 5 | 4 | 3 | 5 | 4 | 4 | 4.17 |
| Yellow | 5 | 4 | 3 | 4 | 4 | 4 | 4.00 |
| White | 4 | 4 | 4 | 4 | 3 | 4 | 3.83 |
| Green | 5 | 4 | 3 | 3 | 5 | 3 | 3.83 |
| Red | 5 | 3 | 2 | 3 | 4 | 3 | 3.33 |
| Default | 3 | 4 | 3 | 3 | 2 | 4 | 3.17 |

**Table A-25** Scorecard for Q05 (Mars Colonization) Across the Six Thinking Dimensions

## *Dimension-by-Dimension Notes (with Evidence Pointers)*

**Black—Risk and Caution (4.67):** Excellent persona fidelity: structured threat taxonomy (health, geopolitical, ethical), cost bands, and a crisp recommendation ("Lunar Genesis" phased path). Strong actionability and distinctiveness. The report could provide more detailed quantification of specific risk likelihoods and mitigation costs, but overall, it is a standout risk brief.

**Blue—Orchestration and Process (4.17):** Clear phase-gated plan (0–12 months; years 2–5; decades), with governance and monitoring (scenario planning, ethics council). Actionability is high; specificity is moderate (technologies listed, fewer KPIs/acceptance criteria). Add decision gates (e.g., TRL thresholds for radiation shielding) to lift SPEC.

**Yellow—Value and Optimism (4.00):** Persuasive case for benefits (spin-offs, collaboration, inspiration) and a phased, sustainable approach. Good balance acknowledging risks; could cite exemplars/metrics for "societal inspiration" and define ROI frames (scientific, economic, diplomatic) to raise SPEC.

**White—Evidence and Facts (3.83):** Balanced feasibility read (propulsion, ISRU, time-to-Mars), risk categories, and trillion-scale cost framing, with prudent phased recommendation. Improves with explicit evidence chains (sourceable figures, uncertainty ranges) and comparative baselines (e.g., lunar ISRU vs. Martian).

**Green—Creativity and Options (3.83):** Inventive proposals (bio-radiation shielding, localized "mini-terraforming," nanobot swarms, cybernetic adaptation). Distinctive and horizon-expanding, but actionability and test plans need tightening (what to prototype where, with what success criteria).

**Red—Affective Intelligence (3.33):** Strong emotional fidelity (dread, caution) and a humane redirect to Earth-first resilience and safer space habitats. To improve SPEC/ACT, translate intuition into stakeholder-impact analyses (crew psych risk models, public support elasticity) and concrete guardrails.

**Default—Baseline (3.17):** Competent survey; limited distinctiveness/persona fidelity. Useful as a neutral reference, but adds little beyond dimensional drafts.

## *Cross-Dimension Synthesis*

Across dimensions, a coherent strategy emerges: (i) *phase before scale* (Blue, Black); (ii) *value with safeguards* (Yellow, White); (iii) *probe creatively, prototype locally* (Green); (iv) *center human limits and ethics* (Red). Convergence points include start lunar/near-Earth to derisk ISRU and closed-loop life support; invest in radiation shielding, autonomy, and psych-safety; use robotic-first build-out; maintain stringent planetary protection; and structure decision gates with abort paths tied to evidence.

## *Actionable Advice and Next Steps*

1. **Define Decision Gates (Blue/White):** Set TRL and metric thresholds for five critical systems: (a) radiation shielding (dose-rate targets), (b) closed-loop life support ($O_2$/$H_2O$ recycle rates), (c) ISRU (kg/day water/propellant extraction), (d) autonomous construction (print speed/MTBF), (e) behavioral health (validated psych-safety indices). Tie funding tranches to gate passage.
2. **Derisk on the Moon and in Orbit (Black/Blue):** Execute a *robotic-first* build of a lunar ISRU demonstration and an orbital bioregenerative life support pilot before any Mars crew scale-up. Use these as kill-or-commit inflexion points.
3. **Embed Ethics and Planetary Protection (Red/White):** Constitute an independent *Mars Ethics & Biosafety Council* with veto power over mission phases; preregister protocols for contamination control and crew well-being thresholds.
4. **Targeted Creative Prototyping (Green):** Fund 3–5 pathfinder experiments: bio-based radiation absorbers, localized pressure-dome "mini-terraformers," regolith-to-structure pipelines, and human-augmentation countermeasure trials—each with 12–24 month lab/analog milestones.
5. **Value Realization Plan (Yellow):** Specify near-term Earth spill-overs (e.g., extreme-environment agriculture, recycling tech, remote autonomous mining), with KPIs and tech-transfer vehicles to justify spend and sustain public support.

## *Summary and Advisory on Evaluation Results*

The evaluation of responses to Question 5 (*Should humanity invest heavily in colonizing Mars within this century?*) demonstrates that each dimension contributed distinctive and complementary perspectives. The **Black Dimension** delivered the strongest overall performance, with rigorous risk analysis, structured categorization of threats, and clear phased recommendations. Its high scores in persona fidelity, actionability, and distinctiveness made it the most robust contribution. The **Blue Dimension** similarly excelled in orchestrating phased processes and governance mechanisms, though it requires more concrete technical thresholds to strengthen specificity.

The **Yellow Dimension** provided a persuasive vision of the potential opportunities and inspirational value of Mars colonization. Its optimism was grounded in benefits such as spin-off technologies, global collaboration, and cultural inspiration, although stronger evidence chains and quantitative ROI frameworks would improve its specificity. The **White Dimension** offered a balanced evidence-based perspective, fairly highlighting feasibility, costs, and risks. However, while its clarity and coverage were solid, its reliance on general figures limited the sharpness of its evidential foundation.

The **Green Dimension** was highly distinctive, generating creative and unconventional proposals such as bio-shielding, nanobot swarms, and mini-terraforming approaches. This innovation added substantial breadth to the discourse but suffered from weaker actionability, as its proposals were not translated into feasible test plans. By contrast, the **Red Dimension** leaned heavily into emotional and ethical cautions, providing authentic human-centered reflections but lacking operational detail. Its relatively lower scores highlight the need to translate affective insights into stakeholder-impact models and practical guardrails.

The **Default Response** functioned as a useful baseline but lacked persona fidelity and distinctiveness compared to the six specialized perspectives. While competent in summarizing the debate, it added limited incremental value. Overall, the interplay of all six dimensions suggests that a sustainable strategy requires convergence: phased and evidence-driven implementation (White, Blue, Black), clear articulation of societal and ethical guardrails (Red), integration of creative pilot projects (Green), and strong communication of tangible benefits (Yellow).

In light of these findings, the recommended path forward is a phased, risk-managed approach beginning with lunar and orbital demonstrations before scaling toward Mars. Decision gates should be defined across five critical systems: radiation shielding, closed-loop life support, in situ resource utilization, autonomous construction, and behavioral health. Creative experiments proposed by the Green Dimension should be tested in controlled analogs to expand the option space. At the same time, the Red and Yellow Dimensions remind us that ethical safeguards and societal inspiration must underpin the endeavor. Taken together, these perspectives show that colonizing Mars should not be viewed as a single leap, but as a disciplined sequence of tested, safeguarded, and value-driven steps.

# Appendix 6

## Question

Should countries introduce a universal basic income (UBI) for all citizens? Discuss financial feasibility, social impact, risks of dependency, potential benefits, and innovative funding models.

## Results

### *Evaluation of Q6—Universal Basic Income (UBI) Across Six Thinking Dimensions*

**Scope:** I evaluated the six dimension-specific responses (White, Red, Black, Yellow, Green, Blue) to: *"Should countries introduce a universal basic income (UBI) for all citizens? Discuss financial feasibility, social impact, risks of dependency, potential benefits, and innovative funding models."* Results below use the requested rubric (PF, COV, SPEC, ACT, DIST, CLAR; 1–5 scale).

### *Scoring Rubric (Applied)*

The **Scoring Rubric (Applied)** evaluation is summarised in **Table A-26 — Scorecard for Evaluation Across the Six Thinking Dimensions with Averages**. **Scale:** 1=poor, 2=fair, 3=good, 4=very good, 5=excellent.



A. F. Vermeulen, *Agentic Hyper-Personalized Dimensions*, https://doi.org/10.1007/979-8-8688-1877-6

| Dimension | PF | COV | SPEC | ACT | DIST | CLAR |
|---|---|---|---|---|---|---|
| White | 5 | 5 | 4 | 4 | 4 | 5 |
| Red | 5 | 3 | 2 | 2 | 4 | 3 |
| Black | 5 | 4 | 3 | 4 | 5 | 4 |
| Yellow | 5 | 4 | 3 | 4 | 4 | 4 |
| Green | 4 | 4 | 2 | 3 | 5 | 3 |
| Blue | 5 | 4 | 3 | 5 | 5 | 4 |
| **Averages** | **4.8** | **4.0** | **2.8** | **3.7** | **4.5** | **3.8** |

**Table A-26** Scorecard for Evaluation Across the Six Thinking Dimensions with Averages

## *Findings by Dimension (Brief Justifications)*

- **White (5/5 PF):** Strong factual framing with explicit cost bands (e.g., global $12–$15T), revenue options (progressive, digital, carbon), and disciplined caveats on labor participation and inflation. Clear recommendations for pilots and phased rollout lift ACT and CLAR. SPEC could rise with country-level numerics and citations to specific pilots.
- **Red (5/5 PF):** Captures affect, intuition, and societal "purpose" risks well (dependency, cohesion); however, limited empirical anchors reduce SPEC and ACT. Tightening claims with evidence (e.g., measured psychological outcomes) would improve COV and CLAR.
- **Black (5/5 PF):** Exemplary risk register (capital flight, labor supply, inflation, moral hazard) and mitigation stance (monitoring, phased trials). Some magnitudes are overstated, trimming SPEC. Still, process-credible and balanced.
- **Yellow (5/5 PF):** Opportunity lens is consistent: entrepreneurship, health, civic engagement, diversified funding (carbon, VAT reform, automation taxes). SPEC is moderate; citing results from named pilots would strengthen it further.
- **Green (4/5 PF):** Creative breadth (robot tax, dynamic payouts, digital assets) and scenario thinking are substantial. A cost line of "$12–$24 per citizen annually" appears erroneous, hurting SPEC/CLAR; fix units and add worked examples.
- **Blue (5/5 PF):** Excellent orchestration: crisp questions, scope control, decision process, and next steps. Cost figures are conservative and sometimes imprecise, but ACT and DIST are powerful due to facilitation quality.

## *Cross-Cutting Strengths and Gaps*

**Strengths:** Persona fidelity is consistently high across all six; risks and opportunities are both addressed (Black vs. Yellow), with a strong facilitation wrapper (Blue) and data-led stance (White).

**Gaps:** Specificity lags (avg. 2.8): several claims need country-calibrated budgets, labor elasticities, and inflation pass-throughs; Green contains a cost unit error; Red needs evidence for psychological and social-cohesion assertions. References to pilots and meta-analyses should be explicitly tied to quantified outcomes across demographics.

## *Actionable Improvements (Next Iteration Checklist)*

1. **Country Packs (Raise SPEC to 4+):** For two contrasting economies (e.g., UK vs. Finland), produce worked UBI budgets at three benefit levels (poverty line, subsistence, partial), with financing mixes (progressive tax, carbon, VAT, automation). Include five-year debt, inflation, and labor-participation sensitivities (White/Blue lead; Black stress-tests.)
2. **Pilot Evidence Table:** Summarize outcomes (employment hours, business formation, health metrics, schooling) from referenced pilots; quantify effect sizes and confidence intervals to back Red/Yellow claims (White curates; Yellow interprets.)
3. **Correct and Calibrate Green:** Fix cost units; provide 2–3 modeled revenue yields for "robot tax," digital asset taxes, and dynamic locale-based payouts with guardrails to limit inflation variability.
4. **Risk Playbooks:** From Black, formalize triggers and countermeasures (e.g., automatic stabilizers that taper UBI growth with CPI-X or labor-force participation dips). Add capital flight and price-level containment protocols.
5. **Red Evidence Bridge:** Pair affective hypotheses (purpose, cohesion) with survey designs and pre/post pilot psychometric instruments to move Red from intuition to testable claims.

## *Advisory Verdict (Research Expert Synthesis)*

The six-dimensional set is decision-ready in *framing* (PF 4.8, DIST 4.5) and *governance/process* (Blue ACT 5), but not yet investment-grade in *specificity* (avg. 2.8). Proceed with a **phased, evidence-seeking path**: design two national country packs, launch tightly scoped pilots with preregistered metrics, wire in Black's stabilizers, and publish White-led technical appendixes. Repair Green's cost units, add Yellow's opportunity KPIs (entrepreneurship, health), and ground Red's claims with validated instruments. This will lift SPEC and CLAR, enabling a balanced, ethically defensible recommendation on UBI feasibility and design.

## *Summary and Advisory on Q06 Evaluation Results*

The evaluation of the six-dimensional responses to Question 06 on the introduction of a universal basic income (UBI) highlights consistent strengths in persona fidelity and balance across perspectives. Each dimension maintained a precise alignment with its intended role, with the White Dimension providing a rigorous evidence base, the Red capturing emotional and intuitive responses, the Black identifying systemic risks, the Yellow emphasizing potential benefits, the Green exploring innovative funding concepts, and the Blue ensuring process orchestration. Overall, the set demonstrates a high degree of complementarity and offers a comprehensive multilens exploration of the UBI debate.

The quantitative rubric indicates strong performance in persona fidelity (average 4.8/5) and distinctiveness (4.5/5), ensuring that each dimension contributed a unique, nonoverlapping perspective. Coverage (4.0/5) was solid, with most dimensions addressing the key remit areas, though the Red and Green outputs showed occasional weaknesses in depth and precision. The most significant shortfall was in specificity (2.8/5), where several dimensions made broad or uncalibrated claims without sufficient empirical grounding, including misstatements of cost scales in the Green Dimension and limited evidence backing psychological assertions in the Red Dimension.

Strengths of the corpus include the White Dimension's grounding in fiscal and social evidence, complemented by the Blue Dimension's facilitation of decision-making processes. The Black Dimension provides a robust catalogue of risks and systemic vulnerabilities, while the Yellow Dimension emphasizes constructive optimism, highlighting pathways to innovation and social value. The Green Dimension contributes creative breadth, introducing unconventional funding models, though its quantitative reliability requires correction. Taken together, the dimensions provide a near-complete framework for evaluating UBI, albeit one that still needs further sharpening of quantitative and evidentiary detail.

Key gaps identified relate to the insufficient specificity of claims and a need for stronger linkage to real-world pilot results and country-level modeling. To address this, further work should focus on developing country packs that provide costed scenarios for diverse economies, with sensitivity analyses for inflation, employment, and fiscal stability. Red's intuitive claims should be paired with survey data and psychometric measures, while Green's speculative funding models must be anchored by tested budgetary mechanisms. Black's risk playbooks should be formalized with predefined triggers and countermeasures, ensuring resilience to unintended macroeconomic consequences.

In conclusion, the advisory recommendation is to advance toward a phased, evidence-seeking path for UBI evaluation and potential implementation. The six-dimensional framework provides a robust strategic baseline but must be supplemented with empirical data, pilot-derived insights, and corrected quantitative models. This dual focus on evidence and process will ensure that UBI evaluations are not only theoretically comprehensive but also practically viable, ethically defensible, and context-sensitive across varied national landscapes.

# Appendix 7

## Question

How would you monetize Dark Data (unknown data without current practical use in business processes)?

## Results

### *Scoring Rubric (Applied)*

The **Scoring Rubric** used for evaluation is presented in **Table A-27 — Scoring Rubric for Evaluation Criteria**.
**Scale:** 1=poor, 2=fair, 3=good, 4=very good, 5=excellent.

| Criterion | Description |
| --- | --- |
| Persona fidelity (PF) | Adherence to the specified role's content and style constraints. |
| Coverage (COV) | Completeness relative to the persona's remit (e.g., risks, facts, opportunities). |
| Specificity (SPEC) | Use of precise, concrete details; avoidance of generic statements. |
| Actionability (ACT) | Presence of clear, feasible next steps where appropriate. |
| Distinctiveness (DIST) | Clear differentiation from other personas' outputs. |
| Clarity (CLAR) | Organization, readability, and length discipline. |

**Table A-27** Scoring Rubric for Evaluation Criteria

## *Quantitative Results (Question 7)*

| Persona | PF | COV | SPEC | ACT | DIST | CLAR | Overall |
|---------|----|----|------|-----|------|------|---------|
| Default | 3 | 4 | 3 | 3 | 2 | 4 | 3.17 |
| White | 4 | 4 | 3 | 3 | 4 | 4 | 3.67 |
| Red | 5 | 3 | 2 | 2 | 5 | 3 | 3.33 |
| Black | 5 | 5 | 4 | 4 | 4 | 4 | 4.33 |
| Yellow | 5 | 5 | 4 | 5 | 4 | 4 | 4.50 |
| Green | 5 | 4 | 3 | 3 | 5 | 3 | 3.83 |
| Blue | 5 | 5 | 4 | 5 | 4 | 4 | 4.50 |

**Table A-28** Quantitative Evaluation Results for Question 7 Across Personas

## *Qualitative Justifications (by Persona)*

**Default**: Broad survey of strategies (licensing, APIs, marketplace) with risks and ethics, but remains generic and least distinctive; actionable steps are present, yet high-level.

**White**: Evidence-first framing and clear definitions; balanced enumeration of avenues and constraints with explicit ethics and a caveat on quantum considerations. Specificity and stepwise guidance are adequate but could be further developed with methods and metrics.

**Red**: Strong persona fidelity via affect, intuition, and "resonance" framing; creative concepts (Sensory Archive, Narrative Resonance, Anomaly Market) but light on concrete safeguards, KPIs, and execution detail.

**Black**: Exemplary risk posture (poisoning, compliance, over-reliance, existential risks) with practical controls (sandboxing, decentralized cohorts, watermarking, quantum-resistant crypto) and continuous monitoring loop; slightly fewer hard metrics but highly operational.

**Yellow**: Opportunity-led, phased roadmap with models, partners, and revenue paths; includes an example impact estimate (15–20% downtime reduction) and ethics-by-design — very actionable and commercially sharp.

**Green**: Inventive portfolio (Neural Seed Licensing, Pattern Resonance Platform, creative "Echoes") and thoughtful risk notes; could translate provocations into clearer delivery plans and validation criteria.

**Blue**: Orchestrates process end-to-end (inventory, workshops, model portfolio, ethics, KPIs, feedback loops); highly actionable facilitation with strong coverage and governance integration.

## *Cross-Dimension Insights*

- **Best Persona Fidelity:** Red/Black/Yellow/Blue align tightly to their roles; White is solid; Default is intentionally neutral.
- **Actionability Leaders:** Yellow and Blue (phased plans, KPIs, monetization levers). Black provides robust controls suitable for immediate adoption.
- **Risk and Ethics Depth:** Black is strongest; White and Blue provide consistent governance; Yellow integrates ethics pragmatically.
- **Creativity:** Green and Red push novel frontiers; Yellow balances creativity with market fit.

## *Targeted Improvements (by Persona)*

**Default**: Add a decision tree mapping data classes to monetization options, with gating checks (lawful basis, DPIA need) and minimal KPIs to lift ACT/DIST.

**White**: Increase SPEC/ACT by naming concrete techniques (e.g., linkage risk measurement, reidentification audits, uncertainty quantification) and a metric pack (precision/recall for anomaly use cases).

**Red**: Preserve tone but append a one-page "ethics and consent" checklist and a micropilot protocol (sample size, debrief, harm screen) to raise ACT/CLAR without diluting persona.

**Black**: Add quantitative thresholds (e.g., acceptable privacy loss $\epsilon$, incident MTTR targets) and red team playbooks to push SPEC/ACT to 5.

**Yellow**: Expand impact ranges beyond a single example (attach baseline vs. uplift tables) and define partner selection criteria to reach SPEC/CLAR=5.

**Green**: Convert provocations into experiment charters (hypothesis, success metric, budget cap, timeline) and a validation cadence to improve ACT/CLAR.

**Blue**: Include a RACI for governance forums and a cadence (e.g., monthly value council, quarterly ethics review) to cement accountability.

## *Recommendation*

Adopt **Blue** to run the program, **Black** to harden controls, and **Yellow** to drive the commercial roadmap. Bring **White** in for evidence standards, and use **Green** and **Red** to seed and test high-variance opportunities under strict, ethical micropilots. This blend maximizes value while safeguarding trust and compliance.

## *Summary and Advisory on Evaluation Results*

The evaluation of responses to Question 7 demonstrates that the six persona dimensions, together with the baseline (Default), provided complementary yet contrasting contributions to the problem of monetizing Dark Data. The quantitative scoring highlights that Yellow and Blue achieved the highest overall ratings (4.50), followed closely by Black (4.33), while White and Green delivered solid but less operationally grounded results. Red maintained excellent persona fidelity but was weaker in specificity and actionability, while Default predictably scored lowest for distinctiveness and depth.

From a strengths perspective, the results clearly illustrate the differentiated value of each persona. Yellow excelled in actionable opportunity framing, with commercially viable roadmaps and measurable benefits, while Blue provided strong orchestration, process structuring, and governance integration. Black distinguished itself by delivering a thorough treatment of risk, compliance, and systemic safeguards, making it the natural anchor for ensuring trust and resilience in Dark Data monetization. White delivered balanced factual clarity, grounding discussions in evidence, but required greater depth in technical specificity. Green and Red offered creative and affective resonance, expanding the conceptual possibility space and capturing the human and imaginative dimensions.

However, the evaluation also revealed areas for targeted improvement. White and Green would benefit from greater translation of concepts into operational steps and more precise metrics, enhancing both actionability and clarity. Red requires the addition of basic frameworks for pilot protocols and ethics to avoid remaining overly abstract. Black, while strong, could push further by defining quantifiable thresholds and response playbooks to make its risk management outputs more measurable. Yellow and Blue, despite their leadership, still require sharper criteria for partner selection, KPIs, and accountability structures to achieve excellence across all dimensions.

Taken together, the personas form a holistic architecture for handling Dark Data: White provides evidence, Red provides emotional resonance, Black mitigates risks, Yellow identifies and drives opportunities, Green injects creativity, and Blue ensures coordination and governance. Their combined insights suggest that no single perspective suffices. Instead, the interplay of structured facilitation, commercial optimism, disciplined caution, and creative exploration allows an enterprise to move responsibly from discovery to monetization. The evaluation confirms that strength lies in balance, with orchestration (Blue) and guardrails (Black) underpinning innovation and opportunity (Yellow, Green, Red) within an evidence-based frame (White).

The advisory conclusion is therefore to operationalize monetization of Dark Data through a council-of-personas model. Practically, this means assigning Blue as the facilitator, Yellow as the growth driver, and Black as the guardian of resilience, while drawing upon White for evidential grounding, Green for experimental creativity, and

Red for intuitive and empathic resonance. This ensemble approach ensures that monetization pathways are commercially valuable, ethically robust, and strategically coordinated. By aligning the dimensions into a managed process, enterprises can convert Dark Data from latent potential into sustainable advantage without compromising trust or compliance.

# Appendix 8

## From Dark Data to Light Data to Monitored Data

### Introduction

Organizations generate vast amounts of data, yet much of it remains hidden in the shadows as *Dark Data*—unused, unstructured, or forgotten. This unused information represents a lost opportunity for insights, efficiency, and innovation. By systematically uncovering Dark Data, transforming it into structured and usable *Light Data*, and embedding it into a continuous monitoring pipeline, businesses can unlock new value streams while improving governance, compliance, and decision-making.

This chapter presents a framework for identifying Dark Data through 25 key questions, aligned with five categories: **People, Process, Technology, Governance, and External**. Each question is linked to a dark-light-monitored transformation pathway, with commentary on why the question is essential for business monetization.

### Framework Questions and Transformation Pathways

#### *People (Human-Generated Data)*

**Q1:** Which collaboration tools (email, chat, documents, recordings) contain valuable context not structured or linked to workflows? **Why It Matters:** Collaboration tools capture real-time decisions, customer feedback, and innovative ideas. Unlocking this information reduces duplication, accelerates onboarding, and creates monetizable knowledge flows. **Dark:** Human communication is stored in unstructured formats.


A. F. Vermeulen, *Agentic Hyper-Personalized Dimensions*, https://doi.org/10.1007/979-8-8688-1877-6

**Light:** Extract, classify, and structure insights via NLP and tagging. **Monitored:** Dashboards track knowledge flows and engagement.

**Q2:** Where are employee skills, certifications, or knowledge repositories stored but not leveraged? **Why It Matters:** Surfacing hidden skills enables more efficient workforce deployment, reduces training spend, and creates opportunities to monetize expertise in new markets. **Dark:** Dormant HR and skills records. **Light:** Structured into knowledge graphs. **Monitored:** Dashboards track skills distribution and gaps.

**Q3:** Which feedback channels (surveys, suggestion boxes, forums) collect input that is never acted upon? **Why It Matters:** Feedback contains signals for product innovation and customer retention. Monetization comes from reducing churn and creating higher-value services aligned with expressed needs. **Dark:** Surveys and feedback logs. **Light:** Categorized insights linked to KPIs. **Monitored:** Dashboards show sentiment and follow-up actions.

**Q4:** What training or onboarding materials are archived but not reused to improve performance? **Why It Matters:** Training costs are high; archived resources represent untapped intellectual capital. Structuring them creates a knowledge base that shortens onboarding and improves productivity. **Dark:** Archived training content. **Light:** Structured learning assets tagged to roles. **Monitored:** Usage and effectiveness dashboards.

**Q5:** Where do recorded meetings, interviews, or presentations exist without structured indexing? **Why It Matters:** Recordings often capture strategic insights and innovation. Indexing them reduces knowledge loss, supports compliance, and creates assets that can drive future revenue. **Dark:** Audio/video recordings. **Light:** Transcripts with metadata. **Monitored:** Searchable repositories with usage tracking.


## *Process (Operational Data)*

**Q1:** Where do manual processes create unstructured or semi-structured data that gets stored but never analyzed? **Why It Matters:** Manual records often reveal bottlenecks and inefficiencies. Structuring them enables process automation, reducing costs and improving throughput. **Dark:** Notes, spreadsheets. **Light:** Standardized formats linked to KPIs. **Monitored:** Monitored usage and relevance scoring.

**Q2:** Which business units generate reports or dashboards that no decision-maker regularly reads? **Why It Matters:** Unused reports waste resources. Identifying and rationalizing them saves money and ensures reporting is tied to monetizable outcomes. **Dark:** Redundant or unused reports. **Light:** Consolidated KPI-linked outputs. **Monitored:** Automated monitoring of report usage.

**Q3:** What workflow exceptions or error logs are recorded but not tracked for root cause analysis? **Why It Matters:** Unmonitored errors create hidden costs and risks. Analyzing exceptions uncovers systemic inefficiencies, reducing downtime and enabling monetizable improvements. **Dark:** Exception/error logs. **Light:** Categorized root causes. **Monitored:** Continuous monitoring of exceptions.

**Q4:** Where do supply chain processes produce redundant paper trails or digital files? **Why It Matters:** Redundant supply chain data hides opportunities for cost optimization and supplier renegotiation. Structuring it can reveal savings and monetizable insights. **Dark:** Paper/digital redundancies. **Light:** Consolidated workflows. **Monitored:** Automated supply chain tracking.

**Q5:** What approval processes generate audit trails that are stored but never reviewed? **Why It Matters:** Approval trails contain compliance insights and negotiation history. Surfacing them ensures accountability and can prevent costly legal exposure. **Dark:** Audit trails. **Light:** Indexed decision points. **Monitored:** Approval compliance dashboards.


## *Technology (System-Generated Data)*

**Q1:** What data is automatically generated by systems, devices, or sensors that no team actively reviews? **Why It Matters:** Sensor and system data often contain early warnings of risks and efficiency gaps. Monetization comes from predictive maintenance and optimized performance. **Dark:** Logs, sensor feeds. **Light:** Structured events. **Monitored:** Integrated into observability platforms.

**Q2:** Which parts of the IT infrastructure generate diagnostic, audit, or monitoring logs that are stored but not converted into insights? **Why It Matters:** Ignored logs mean missed opportunities for cost savings, performance tuning, and threat detection. Monetization comes from reduced downtime and risk. **Dark:** Audit/system logs. **Light:** Structured anomaly detection. **Monitored:** Predictive monitoring with alerts.

**Q3:** Which cloud usage or licensing logs are archived without cost optimization analysis? **Why It Matters:** Unused cloud data can reveal cost inefficiencies and opportunities to renegotiate contracts. Savings go straight to the bottom line. **Dark:** Licensing logs. **Light:** Cost-optimized reports. **Monitored:** Usage dashboards with anomaly detection.

**Q4:** Where do application crash dumps or error traces exist without structured resolution tracking? **Why It Matters:** Crash dumps contain signals about reliability and quality. Analyzing them reduces defects, enhances customer satisfaction, and protects revenue. **Dark:** Application crash data. **Light:** Structured error categories. **Monitored:** Error trend dashboards.

**Q5:** What telemetry from IoT devices is stored but not leveraged for predictive insights? **Why It Matters:** IoT telemetry enables predictive services. Without analysis, opportunities for monetizing proactive maintenance and performance optimization are lost. **Dark:** IoT device data. **Light:** Categorized telemetry streams. **Monitored:** Predictive IoT dashboards.

### *Governance (Compliance and Archives)*

**Q1:** Which datasets are collected "just in case" but not directly linked to business processes or KPIs? **Why It Matters:** "Just in case" data consumes storage costs but may hold compliance or strategic value. Monetization comes from repurposing or eliminating waste. **Dark:** Compliance-driven storage. **Light:** Classified for value and need. **Monitored:** Life cycle management with access tracking.

**Q2:** What historical data exists in archives, backups, or cold storage that is never surfaced for analysis? **Why It Matters:** Historical data offers trends and insights for forecasting. Monetization comes from uncovering long-term customer or operational patterns. **Dark:** Legacy archives. **Light:** Classified for relevance. **Monitored:** Analytics integration.

**Q3:** Where do data retention policies create silos of expired or redundant data? **Why It Matters:** Misapplied retention rules waste resources and obscure valuable insights. Proper governance reduces risk and frees resources for monetizable analysis. **Dark:** Expired/redundant data. **Light:** Policy-aligned datasets. **Monitored:** Monitored data life cycle.

**Q4:** What compliance documents are scanned and stored without text extraction or indexing? **Why It Matters:** Nonsearchable documents reduce regulatory efficiency. Monetization comes from reducing compliance effort and avoiding penalties. **Dark:** Scanned compliance docs. **Light:** Extracted searchable content. **Monitored:** Access-tracked repositories.

**Q5:** Which risk or audit findings are documented but never followed up with metrics? **Why It Matters:** Ignored findings create financial exposure. Analyzing them reduces risk and protects revenue streams. **Dark:** Risk/audit reports. **Light:** Categorized metrics. **Monitored:** Follow-up dashboards.

### *External (Partner and Supplier Data)*

**Q1:** What raw data do external systems send us that we store without further processing? **Why It Matters:** Partner and supplier data often contain hidden efficiency or revenue opportunities. Structuring it strengthens business relationships and monetization potential. **Dark:** Partner/EDI feeds. **Light:** Normalized and structured. **Monitored:** Lineage and freshness tracking.

**Q2:** Where do we have duplicate data sources or shadow databases that are rarely queried? **Why It Matters:** Duplicates waste resources and introduce risk. Consolidation reduces costs and ensures data supports monetizable outcomes. **Dark:** Redundant systems. **Light:** Deduplicated and aligned. **Monitored:** Integrated quality pipelines.

**Q3:** Which vendor invoices, contracts, or communications are stored without structured analysis? **Why It Matters:** Contracts and invoices hold opportunities for renegotiation, compliance savings, and cost optimization-direct monetization opportunities. **Dark:** Contracts and invoices. **Light:** Structured contract analytics. **Monitored:** Compliance monitoring dashboards.

**Q4:** What customer service tickets or partner escalations are archived without trend reporting? **Why It Matters:** Tickets reveal service quality issues. Analyzing them improves retention, reduces costs, and supports monetizable product improvements. **Dark:** Customer/partner tickets. **Light:** Categorized issues. **Monitored:** Escalation trend dashboards.

**Q5:** Where do third-party research or market feeds get saved but not integrated into decision models? **Why It Matters:** Market feeds offer a competitive advantage when integrated. Ignoring them wastes licensing spend and forfeits monetization opportunities. **Dark:** Market/third-party data. **Light:** Integrated insights. **Monitored:** Usage dashboards with ROI tracking.

## Conclusion

Transforming Dark Data into actionable insights requires a structured pipeline. By asking the right discovery questions across people, process, technology, governance, and external data categories, organizations can identify hidden sources of information. Turning these into Light Data ensures they become structured, relevant, and linked to business value. Embedding them into monitored pipelines closes the loop, ensuring continuous oversight, compliance, and operational advantage.

The framework presented in this chapter provides both the *diagnostic lens* to surface Dark Data and the *action pathway* to integrate it into enterprise intelligence, ensuring data serves as a living, monetized asset rather than a forgotten liability.

# Appendix 9

## Building a Profit Playbook for SMEs

### *Introduction*

For small- and medium-sized enterprises (SMEs), achieving profitability signals not just financial success but also indicates sustainability, competitiveness, and long-term viability.

Each department, regardless of its direct involvement in revenue generation or its support role, should address a fundamental question: *"In what ways does this function enhance profit?"*

SMEs often mistakenly view areas like HR, IT, or Legal as mere cost centers, concentrating predominantly on Sales or Finance for profit. This perspective is restrictive. Instead, each department should be considered a potential source of profit, bearing clear responsibilities regarding decisions that affect profit margins, growth, and customer lifetime value.

The following framework presents **ten guiding questions per department**, forming a **Profit Playbook** that leaders can employ to align strategy, evaluate results, and optimize value creation throughout the organization.

### *Sales*

1. **How can we increase conversion rates from leads to paying customers?**
   Improving conversion rates directly transforms marketing spend and lead generation efforts into revenue. Each percentage point increase in conversion means more value extracted from the same input. Analyzing patterns behind successful

deals uncovers insights into customer behavior, enabling the business to design repeatable and scalable revenue strategies.

2. **Which customer segments generate the highest lifetime value (LTV)?**
Not all customers are equally profitable. By identifying high-LTV segments, businesses can focus sales and marketing resources on the most lucrative relationships. This insight allows for targeted product bundling, customer retention efforts, and predictive growth modeling, ensuring that scarce resources generate the greatest financial return.

3. **How can we shorten the sales cycle without losing quality?**
Long sales cycles tie up resources and delay revenue realization. Reducing friction points accelerates cash flow and increases deal volume capacity. Insights into where deals stall enable the business to refine engagement strategies, prioritize quick wins, and standardize efficient processes without sacrificing deal quality.

4. **What upsell or cross-sell opportunities exist to grow revenue per customer?**
Expanding revenue within existing accounts is typically more cost-effective than acquiring new customers. Understanding buying patterns allows businesses to package complementary offerings, maximize account profitability, and build deeper customer relationships. These insights monetize existing trust and reduce acquisition costs.

5. **Which sales activities yield the highest ROI and should be scaled?**
Sales teams often invest in numerous activities, but only some deliver meaningful returns. Measuring ROI at the activity level highlights which tactics (e.g., demos, events, or outbound campaigns) drive the most conversions. Scaling high-ROI activities maximizes profitability while eliminating wasted effort.

6. **Where are pipeline velocity bottlenecks, and how do we remove them?**
Every delay in the pipeline slows down revenue generation. Identifying bottlenecks—whether in qualification, proposal, or negotiation—ensures sales teams spend time where it matters most. Removing these friction points improves throughput and creates a more predictable revenue stream.

7. **What pricing and discounting guardrails maximize margin without hurting win rates?**
Discounts can win deals but often erode profitability. Establishing guardrails ensures consistency, protects margins, and provides clarity to sales teams. Data-driven pricing insights also support dynamic pricing models that align with customer value perception, ensuring revenue growth without margin leakage.

8. **Which channels or partners add the most profitable revenue, and how do we expand them?**
Not all channels contribute equally to profitability. Some drive volume but at low margins, while others deliver fewer but more profitable deals. By tracking partner and channel performance, businesses can double down on high-margin collaborations and exit underperforming ones, directly improving bottom-line results.

9. **How should we prioritize territories and key accounts to focus effort where profit is highest?**
Sales capacity is finite, and spreading it too thin reduces effectiveness. By prioritizing territories and accounts based on revenue potential, customer fit, and competitive positioning, businesses can allocate resources more strategically. This focus increases win rates and ensures that insights directly translate into maximized profitability.

10. **What do win/loss analyses reveal about objections, and how do we operationalize responses?**
Every lost deal represents both a missed revenue opportunity and valuable insight. Systematically capturing and analyzing objections enables businesses to refine messaging, strengthen proposals, and develop counter-strategies. Turning this feedback loop into operational playbooks converts loss data into competitive advantage and future revenue.

## *Finance and Accounting*

1. **How can we improve cash flow and reduce days sales outstanding (DSO)?**
Strong cash flow is the lifeblood of any business. Reducing DSO ensures quicker access to capital, lowering reliance on external financing and enabling reinvestment into growth. Insights into payment behaviors and credit policies help monetize working capital by keeping funds moving efficiently through the business.

2. **Which costs are eroding margins, and how can we optimize them?**
Hidden or poorly managed costs can significantly impact profitability. By identifying and analyzing cost drivers, businesses can focus on efficiency initiatives, renegotiate supplier contracts, or streamline operations. These insights directly increase margin retention and long-term sustainability.

3. **What is our break-even point, and how do we stay above it?**
The break-even point defines the minimum revenue required to cover costs. Understanding and managing this threshold allows businesses to plan pricing, sales volumes, and expense control with confidence. Consistently operating above break-even ensures profitability and reduces vulnerability to downturns.

4. **Which revenue streams deliver the best profitability ratios?**
Not all revenues are equal—some streams generate far higher returns than others. Analyzing profitability ratios by segment reveals which offerings deliver the best margins, helping businesses prioritize profitable lines while reevaluating underperforming ones. This maximizes return on effort and investment.

5. **How can financial forecasting help prevent risks and capture opportunities?**
Forecasting equips decision-makers with foresight into revenue, expenses, and market fluctuations. By modeling best- and worst-case scenarios, businesses can avoid liquidity crises and capitalize on growth opportunities ahead of competitors. This turns financial insights into proactive monetization strategies.

6. **What are our unit economics (e.g., LTV/CAC, gross margin per product) and how do we improve them?**

   Unit economics measure the profitability of individual products, customers, or services. Strong unit economics indicate a scalable business model. Tracking metrics like LTV/CAC ensures that growth translates into profitability, not just revenue expansion, guiding resource allocation toward high-return areas.

7. **What working capital levers (inventory, payables, receivables) free cash without hurting growth?**

   Optimizing working capital allows businesses to unlock trapped cash. For example, reducing inventory cycles, extending payables where possible, and accelerating receivables improve liquidity. These levers increase available capital for reinvestment while supporting sustainable growth.

8. **How should we govern pricing, discounting, and margin leakage across quotes and contracts?**

   Poorly governed pricing practices can erode margins silently. Establishing strong financial governance ensures that discounts and quotes remain profitable while maintaining competitiveness. Monitoring leakage across contracts allows businesses to enforce discipline and protect profitability systematically.

9. **What is the true cost-to-serve by segment or product, and where should we reprice or exit?**

   Some segments or products consume disproportionate resources relative to the value they generate. By calculating the true cost-to-serve, businesses uncover which areas to optimize, reprice, or even exit. This ensures resources are reallocated toward offerings that create higher profitability.

10. **Which internal controls reduce leakage, fraud, and error while keeping overhead lean?**

    Effective controls protect revenue while minimizing waste, fraud, and inefficiency. Striking the balance between strong oversight and lean operations ensures that profitability is safeguarded without burdening the business with excessive bureaucracy or costs.

## *Operations*

1. **How can we reduce production or service delivery costs without lowering quality?**

   Cost reduction in operations must not compromise quality, as customer satisfaction directly drives repeat revenue. By analyzing inefficiencies, waste, and resource allocation, businesses can identify savings that improve margins. Smart cost optimization ensures profitability while protecting brand reputation.

2. **Where are bottlenecks slowing down throughput or customer fulfilment?**

   Bottlenecks in production or fulfilment delay revenue recognition and damage customer trust. Identifying and removing these constraints enables smoother workflows, faster order completion, and higher capacity utilization. Insights into bottlenecks directly translate into higher revenue per unit of time.

3. **What process improvements deliver the best efficiency gains?**
Process optimization reduces waste and increases productivity, which lowers costs per unit delivered. Prioritizing improvements that yield the most significant efficiency gains ensures businesses capture maximum return on investment from operational initiatives, turning efficiency into profit.

4. **How can we improve quality control to reduce rework and warranty costs?**
Defects, rework, and warranty claims erode margins and customer trust. Strong quality control minimizes these costs, reduces customer churn, and improves brand equity. Insights into defect trends allow businesses to monetize operational excellence by lowering hidden costs.

5. **How can we scale operations flexibly as demand increases?**
Scalability ensures that businesses can grow revenue without proportionally increasing costs. Flexible operations—supported by automation, modular processes, and scalable supply chains—enable rapid response to demand while preserving profitability margins.

6. **Which activities should be made, bought, or partnered to optimize cost and speed?**
Make-vs.-buy decisions shape operational cost structures. Partnering strategically can reduce fixed costs and increase speed to market, while in-house control may protect intellectual property or margins. Data-driven decisions here ensure an optimal balance between price, speed, and profitability.

7. **How do we improve capacity planning (S&OP) to cut stockouts and overproduction?**
Poor planning leads to lost sales from stockouts or wasted capital from overproduction. Improving Sales & Operations Planning (S&OP) ensures accurate demand forecasting, reducing working capital strain while maximizing revenue capture. Effective planning converts insights into both efficiency and profitability.

8. **What inventory policies (turns, safety stock) minimize working capital and obsolescence?**
Excess inventory ties up cash and risks obsolescence, while insufficient stock leads to lost sales. Balancing safety stock and inventory turns ensures liquidity is protected and customer demand is met. Insights here monetize operational agility by freeing trapped capital.

9. **Which suppliers deliver the best total cost of ownership, and how do we improve their performance?**
Supplier costs extend beyond purchase price—quality, delivery reliability, and life cycle costs all matter. Evaluating suppliers on the total cost of ownership ensures procurement drives profitability. Strengthening supplier performance turns partnerships into a competitive advantage.

10. **How do safety and compliance improvements reduce downtime and hidden costs?**
Unsafe or noncompliant operations risk fines, accidents, and costly downtime.

Proactive investment in safety and compliance reduces these risks while boosting employee productivity and morale. Insights here safeguard both revenue and reputation by preventing avoidable losses.

## *Marketing and Communications*

1. **Which channels deliver the lowest customer acquisition cost (CAC)?**
   Understanding which channels generate customers at the lowest CAC allows marketing spend to be directed where it creates the most efficient growth. By focusing on profitable channels, businesses can scale acquisition without over-spending, ensuring that each marketing dollar drives maximum margin.
2. **How can we increase brand awareness in target profitable segments?**
   Awareness in the right segments attracts high-value customers and shortens the sales cycle. By aligning awareness campaigns with profitability insights, businesses can ensure that visibility translates directly into revenue growth rather than wasted reach.
3. **What marketing messages resonate best to drive conversions?**
   Tailoring messages to customer motivations increases engagement and purchase likelihood. Data-driven insights into language, tone, and value propositions ensure that communications convert prospects efficiently, maximizing return on marketing investment.
4. **How can we track ROI on every campaign and optimize spend?**
   Campaigns without measurable ROI risk draining resources. Tracking and ana-lyzing performance ensures money is spent only on initiatives that drive conver-sions and revenue. Continuous optimization turns marketing into a predictable and profitable engine.
5. **How can we turn satisfied customers into advocates and referrals?**
   Customer advocacy reduces acquisition costs and increases trust among prospects. Turning positive experiences into structured referral programs mon-etizes satisfaction by leveraging existing relationships to generate new, low-cost revenue streams.
6. **What positioning or category narrative allows premium pricing and faster adoption?**
   Strong positioning supports both differentiation and pricing power. By shaping the narrative of a category, businesses can justify higher margins and encourage faster adoption, ensuring that brand equity translates into tangible financial value.
7. **Which content and offers most effectively move prospects between funnel stages?**
   Matching content to each stage of the buyer journey ensures smoother progres-sion from awareness to conversion. Insights here prevent drop-offs, improve pipeline velocity, and increase total conversion rates, monetizing customer en-gagement at each step.

8. **What is the optimal lead quality threshold to protect sales productivity and margins?**
   Poor-quality leads waste sales resources and lower win rates. Defining and enforcing a lead quality threshold ensures sales teams focus on the most profitable opportunities, aligning marketing outputs with revenue objectives and protecting overall profitability.

9. **Which retention and loyalty programs have positive unit economics?**
   Retention costs are typically lower than acquisition costs. Evaluating loyalty initiatives through unit economics ensures programs strengthen customer lifetime value while maintaining profitability, turning retention into a high-margin growth driver.

10. **How do we measure brand health (e.g., NPS, share of voice) and tie it to revenue outcomes?**
    Brand health metrics provide early indicators of long-term revenue potential. By linking measures like NPS and share of voice to actual revenue outcomes, businesses can monetize brand strength, making it a quantifiable and manageable driver of profitability.

## *Customer Service/Support*

1. **How can we reduce churn and increase customer retention rates?**
   Retention directly impacts profitability since keeping a customer is cheaper than acquiring a new one. Reducing churn improves lifetime value and stabilizes revenue, turning customer loyalty into a predictable financial asset.

2. **What issues cost us the most in support time or refunds, and how do we prevent them?**
   High refund or support costs erode profit margins. Analyzing the root causes allows proactive fixes that reduce wasted resources, monetizing efficiency by cutting unnecessary costs.

3. **How can we improve customer satisfaction to boost repeat purchases?**
   Satisfied customers are more likely to return and buy again. Turning satisfaction into repeat business compounds revenue and lowers CAC, strengthening overall profitability.

4. **How can we turn complaints into upsell opportunities?**
   A complaint represents both risk and opportunity. When resolved effectively, it can lead to upsell or cross-sell moments, converting dissatisfaction into additional revenue.

5. **What feedback loops help improve products, services, and profitability?**
   Customer insights reveal unmet needs and opportunities for improvement. Structured feedback loops ensure support data drives innovation, reducing churn and enhancing monetizable value.

6. **Which self-service capabilities (FAQ, portals, chat) deflect tickets at acceptable CSAT?**
   Self-service reduces cost-to-serve while maintaining satisfaction. This creates scalability in support operations while preserving profitability margins.
7. **How do we improve first-contact resolution and SLA design to cut cost-to-serve?**
   First-contact resolution reduces repeat tickets and lowers service costs. Effective SLA design ensures efficiency while protecting revenue from churn caused by poor service.
8. **Which knowledge-base improvements reduce handle time and training costs?**
   Better knowledge bases increase agent productivity, reduce onboarding time, and speed resolution. These improvements directly lower service costs while protecting satisfaction.
9. **What proactive outreach (health checks, CSM plays) prevents churn in at-risk segments?**
   Proactive engagement prevents cancellations before they occur. This protects revenue and ensures customers generate their full lifetime value.
10. **How do we accelerate onboarding time-to-value to expand accounts earlier?**
    Faster onboarding increases early adoption, enabling upsell and expansion sooner. This compresses the revenue realization timeline and maximizes account profitability.

## *Executive Leadership / Strategy*

1. **What markets or niches offer the highest profitable growth potential?**
   Market prioritization ensures resources flow toward the most lucrative opportunities, converting strategic focus into financial growth.
2. **How can we differentiate sustainably from competitors?**
   Differentiation protects pricing power and margin. Sustainable uniqueness transforms competitive positioning into long-term monetization.
3. **How should we allocate resources for maximum ROI?**
   Capital and talent allocation drive profit. Data-backed allocation ensures every investment generates measurable financial returns.
4. **What risks could threaten profitability, and how do we mitigate them?**
   Identifying risks protects revenue streams from erosion. Mitigation ensures business resilience and protects shareholder value.
5. **How do we align all departments to work toward profit-driven goals?**
   Alignment avoids wasted effort and duplication. When all teams drive toward financial outcomes, profit grows systematically.

6. **Which company-wide OKRs most directly improve the P&L, and how will we measure them?**
   Financially aligned OKRs ensure measurement systems track profit impact, creating transparency and accountability.
7. **Where should we build, buy, or partner to accelerate growth and reduce risk?**
   Strategic decisions on build vs. buy vs. partner shape cost efficiency and speed to market, directly impacting ROI.
8. **Is our organization's design clear on accountability, speed of decision, and cost efficiency?**
   Clear structures reduce wasted time, prevent duplication, and improve cost efficiency, monetizing agility in execution.
9. **What operating rhythm (reviews, dashboards, cadences) keeps focus on leading indicators?**
   A disciplined operating rhythm ensures early signals guide profitable decisions before lagging indicators surface.
10. **Which scenarios (best/base/worst) inform capital allocation and optionality?**
    Scenario planning ensures capital is preserved in downturns and maximized in growth cycles, monetizing foresight.


## *Human Resources (HR)*

1. **How can we hire people who directly contribute to profit growth?**
   Strategic hiring ensures headcount drives measurable business outcomes, aligning talent with profitability.
2. **How can training improve productivity and reduce costly mistakes?**
   Training reduces error rates, improves efficiency, and ensures employees add higher value per hour worked.
3. **What motivates employees to perform at their highest profit-impacting level?**
   Motivation strategies maximize output and engagement, ensuring workforce energy translates into profit.
4. **How do we reduce staff turnover to avoid replacement and training costs?**
   Lower turnover reduces recruitment and training expense, protecting both margins and institutional knowledge.
5. **How can performance incentives be aligned with profitability goals?**
   Linking rewards to profit ensures behaviors and outcomes directly reinforce financial performance.
6. **What workforce plan (headcount, skills, span of control) supports growth at the lowest cost?**
   Workforce planning ensures the business scales with efficiency, balancing cost with capability.

7. **How do we strengthen our employer value proposition to reduce time-to-hire and salary premiums?**
   Strong EVPs lower hiring costs and speed recruitment, monetizing employer brand equity.
8. **Which performance management practices raise output without adding bureaucracy?**
   Effective performance management maximizes productivity without burdening the business with overhead.
9. **Where are compliance risks (contracts, overtime, leave) that could incur penalties?**
   Addressing compliance risks early prevents costly fines, lawsuits, and disruptions that harm profitability.
10. **Which cultural behaviors most correlate with high performance, and how do we reinforce them?**
    High-performance cultures ensure consistent output and innovation, monetizing organizational behavior.

## *Information Technology (IT)/Digital Systems*

1. **How can automation reduce costs and increase efficiency?**
   Automation cuts labor costs and increases throughput, converting technology investment into savings.
2. **Which technologies provide the best ROI for scaling operations?**
   Evaluating ROI ensures only tools with intense payback periods drive growth and profitability.
3. **How do we ensure cybersecurity to prevent costly breaches?**
   Strong security protects against losses, fines, and reputational damage, preserving both profit and trust.
4. **How can we use data analytics to identify new profit opportunities?**
   Analytics turns raw data into insights that reveal new markets, products, and cost-saving opportunities.
5. **How do we support remote work and collaboration without extra cost overhead?**
   Smart digital tools reduce real estate and travel costs while maintaining productivity, boosting margins.
6. **What is the total cost of ownership (build vs. buy, life cycle) of our core systems?**
   Understanding life cycle costs prevents overspending and ensures long-term profitability of tech choices.
7. **How do we strengthen data governance and privacy to avoid fines and rework?**
   Compliance with privacy laws avoids penalties and builds trust, monetizing responsible data practices.

8. **Which integrations/APIs remove manual work and errors between systems?**
Integration eliminates duplication and human error, reducing costs and improving efficiency.

9. **What reliability targets (uptime, MTTR) protect revenue and reduce disruption?**
High availability protects customer experience and prevents lost sales from downtime.

10. **How do we manage vendors to avoid lock-in and negotiate better commercial terms?**
Effective vendor management reduces cost exposure and strengthens bargaining power, improving profitability.

## *Legal and Compliance*

1. **How do we minimize legal risks that could create financial losses?**
Proactively managing legal risk prevents costly litigation and fines, protecting profitability.

2. **How can contracts be structured to secure better profit margins?**
Smart contract design improves margins by reducing leakage and protecting revenue.

3. **What regulatory compliance is required to avoid fines or penalties?**
Compliance ensures the business avoids penalties and reputational damage, preserving profit.

4. **How can intellectual property be protected and monetized?**
IP protection safeguards innovation while monetization strategies generate new revenue streams.

5. **Where can negotiation reduce liabilities and increase profitability?**
Effective negotiation minimizes risk exposure while strengthening the financial outcome of deals.

6. **What is our exposure to litigation, and do insurance and reserves cover it prudently?**
Understanding exposure prevents unexpected shocks that could drain profit reserves.

7. **Which competition, export, or sector rules could constrain strategy or pricing?**
Awareness of restrictions allows proactive planning, monetizing compliance agility.

8. **Are our data processing agreements and retention policies aligned to risk and cost?**
Optimized policies balance compliance with efficiency, ensuring legal obligations don't erode profit.

9. **Which compliance processes can be automated to cut cost and error rates?**
   Automation lowers compliance overhead while improving accuracy, preserving margins.
10. **How do we shorten the dispute resolution cycle time and recover cash faster?**
   Faster resolution protects liquidity and ensures capital isn't tied up unnecessarily.


### *Research and Development (R&D)/Innovation*

1. **What unmet customer needs can we solve profitably?**
   Identifying unmet needs ensures R&D spend translates into new revenue streams.
2. **Which innovations deliver the highest ROI within 12–36 months?**
   Prioritizing near-term ROI ensures investments improve profitability on a predictable horizon.
3. **How can we reduce the cost and risk of new product development?**
   Lowering R&D risk ensures higher return rates and prevents wasted capital.
4. **How can we differentiate products to command higher margins?**
   Differentiated products justify premium pricing, monetizing innovation directly.
5. **Which partnerships accelerate innovation without increasing overhead?**
   Collaborations share risk and cost, accelerating returns while protecting margins.
6. **What stage-gate and kill criteria keep investment focused on winners?**
   Disciplined governance ensures poor investments don't drain profitability.
7. **How do we embed customer cocreation and voice-of-customer into roadmaps?**
   Involving customers early ensures innovations meet market demand, maximizing adoption and ROI.
8. **What IP strategy (patents, trade secrets, FTO) best protects returns?**
   Strong IP strategies secure competitive advantage and monetize innovation.
9. **How can toolchains, prototyping, and reuse accelerate time-to-market?**
   Faster launches reduce time-to-revenue and extend product life cycle profitability.
10. **Which postlaunch metrics (adoption, margin uplift) guide iteration and scale?**
   Monitoring postlaunch ensures products evolve profitably and scale effectively.


## Conclusion

The Profit Playbook is not a theoretical exercise. It is a practical framework for turning every function into a profit contributor. For SMEs, where resources are limited and competition is fierce, aligning all departments to a common profitability agenda is essential.

By asking these **ten** questions consistently, each department identifies inefficiencies, unlocks growth opportunities, and ensures that every decision contributes to financial sustainability. This creates a culture where profit is not the sole responsibility of sales or finance, but the shared outcome of coordinated efforts across the enterprise.

Running a profit playbook requires discipline: measure outcomes, adjust strategies, and revisit these questions regularly as the business grows. In doing so, SMEs transform complexity into clarity, aligning daily operations with long-term profitability and resilience.

# Appendix 10

## Business Cases for the Six Thinking Dimensions

### *AIOps*

**Business Problem:** Modern enterprises run highly distributed IT systems that generate massive volumes of telemetry, logs, and alerts. Traditional monitoring tools struggle with alert fatigue, siloed data, and slow incident resolution, leading to high downtime costs and missed optimization opportunities.

**Six Thinking Dimensions:**

- **White (Facts)**: Correlates telemetry, logs, and metrics into structured, accurate insights.
- **Red (Emotion)**: Captures operator sentiment and user experience signals to prioritize incidents.
- **Black (Risk)**: Assesses the risk of outages, cascading failures, and compliance breaches.
- **Yellow (Value)**: Highlights cost savings from predictive maintenance and SLA optimization.
- **Green (Creativity)**: Generates innovative remediation playbooks and anomaly patterns.
- **Blue (Orchestration)**: Governs multiagent swarms to ensure unified root cause analysis.

**Monetization Gains:** Reduced downtime (saving millions in lost revenue), lowered IT operations costs, SLA compliance credits, and new managed AIOps service offerings

## Healthcare

**Business Problem:** Healthcare systems face challenges with fragmented patient data, diagnostic errors, and inefficient workflows. Clinicians often spend more time on administration than on patient care.

**Six Thinking Dimensions:**

- **White**: Integrates EHR, imaging, and lab results into a unified knowledge base.
- **Red**: Aligns treatment plans with patient emotions, preferences, and empathy-driven care.
- **Black**: Identifies risks in medication safety, treatment pathways, and ethical AI use.
- **Yellow**: Quantifies value through improved outcomes, reduced readmissions, and cost efficiency.
- **Green**: Suggests novel therapies, drug repurposing, and personalized treatment models.
- **Blue**: Orchestrates multiagent decision support across care teams and hospital systems.

**Monetization Gains:** Improved patient throughput, reduced malpractice costs, AI-driven drug discovery, and monetization of anonymized health insights.

## Finance

**Business Problem:** Banks and financial institutions face fraud, regulatory complexity, and high competition in digital services. Dark Data in transaction logs and customer records remains underexploited.

**Six Thinking Dimensions:**

- **White**: Provides accurate transaction validation and compliance reporting.
- **Red**: Detects sentiment in customer behavior for proactive service.
- **Black**: Identifies systemic risk, fraud patterns, and credit exposure.
- **Yellow**: Prioritizes profitable customers, investment opportunities, and operational savings.
- **Green**: Designs innovative products (microloans, digital wallets, robo-advisors).
- **Blue**: Coordinates agent councils for fraud detection, compliance, and personalized advice.

**Monetization Gains:** Fraud loss reduction, new financial products, higher customer retention, and regulatory cost savings.

## Public Sector

**Business Problem:** Governments operate complex systems for taxation, defense, and policy, but face inefficiencies, legacy systems, and limited transparency.

**Six Thinking Dimensions:**

- **White**: Aggregates census, economic, and policy datasets into coherent insights.
- **Red**: Captures citizen trust, feedback, and social sentiment in policymaking.
- **Black**: Highlights risks in national security, infrastructure, and compliance.
- **Yellow**: Identifies areas for cost savings and economic growth initiatives.
- **Green**: Suggests innovative governance models and public-private collaborations.
- **Blue**: Orchestrates interdepartmental AI agents for transparency and accountability.

**Monetization Gains:** Reduced administrative overhead, improved tax collection, efficient public spending, and increased citizen satisfaction.

## Citizen Services

**Business Problem:** Citizen-facing services often suffer from poor accessibility, long queues, and fragmented digital experiences, particularly in areas such as welfare, permits, and benefits.

**Six Thinking Dimensions:**

- **White**: Ensures accurate citizen data integration across government services.
- **Red**: Incorporates empathy in service delivery, focusing on vulnerable groups.
- **Black**: Identifies fraud in benefits, identity theft, and systemic abuse.
- **Yellow**: Unlocks value through efficient service delivery and citizen self-service portals.
- **Green**: Develops new digital citizen engagement channels and AI chatbots.
- **Blue**: Coordinates AI agents across departments to deliver a unified "one-stop service."

**Monetization Gains:** Reduced operational costs, higher citizen satisfaction scores, and new revenue models through digital service marketplaces.

## Space Exploration

**Business Problem:** Space agencies and private companies face astronomical costs, vast amounts of data from satellites/telescopes, and high mission risks. Many insights remain buried in unprocessed telemetry and research data.

**Six Thinking Dimensions:**

- **White**: Structures raw sensor data from satellites, rovers, and probes.
- **Red**: Models astronaut well-being and human-machine trust in missions.
- **Black**: Manages mission-critical risks such as system failures and orbital debris.
- **Yellow**: Demonstrates ROI through space resource utilization and spin-off technologies.
- **Green**: Designs novel mission strategies, autonomous navigation, and terraforming concepts.
- **Blue**: Orchestrates agent swarms for mission planning, coordination, and interstellar data relay.

**Monetization Gains:** Commercialization of satellite data, asteroid mining opportunities, technology spin-offs, and public-private partnerships.

# Appendix 11

**Appendix: United Nations Sustainable Development Goals (SDG) Business Cases Empowered for Data for Good**

## *Introduction to the Sustainable Development Goals*

The 17 Sustainable Development Goals (SDGs) form a universal blueprint adopted by the United Nations in 2015 to guide global development through 2030. Framed within the 2030 Agenda, the SDGs aim to deliver lasting peace, prosperity, and planetary health by addressing the world's most urgent challenges—from eliminating poverty and hunger to fostering quality education, robust institutions, and climate resilience. Interconnected and indivisible, these goals require inclusive, multisectoral collaboration and innovative thinking to fulfil their promise of "leaving no one behind" while balancing economic, social, and environmental sustainability.

## SDG 1: No Poverty

**Business Problem:** Rapid population growth strains social safety nets and traps communities in intergenerational poverty. **Six Thinking Dimensions:**

- **White:** Integrates demographic and income data to identify poverty hotspots.
- **Red:** Understands household stress and emotional drivers of poverty.
- **Black:** Assesses risks of population rise on social stability.
- **Yellow:** Highlights the value of targeted family planning and livelihood programs.
- **Green:** Designs community-led income innovation models.
- **Blue:** Orchestrates agents across NGOs, governments, and communities.

© The Author(s), under exclusive license to APress Media, LLC, part of Springer Nature 2026
A. F. Vermeulen, *Agentic Hyper-Personalized Dimensions*,
https://doi.org/10.1007/979-8-8688-1877-6

**Monetization Gains:** Cost savings via efficient aid targeting, reduced welfare burden, and new microfinance models.

## SDG 2: Zero Hunger

**Business Problem:** Growing populations amplify food insecurity and pressure on arable land. **Six Thinking Dimensions:**

- **White:** Maps population density vs. food supply data.
- **Red:** Captures local food demand sentiment and cultural preferences.
- **Black:** Models risks of famine and agricultural collapse.
- **Yellow:** Values investments in sustainable intensification and family planning.
- **Green:** Invents precision agriculture and agroforestry solutions.
- **Blue:** Coordinates agents across farms, distribution systems, and policy.

**Monetization Gains:** Yield improvements, reduced waste, and scalable agritech services.

## SDG 3: Good Health and Well-being

**Business Problem:** High fertility and unintended pregnancies strain healthcare; maternal and child mortality remain high. **Six Thinking Dimensions:**

- **White:** Consolidates health, fertility, and service usage data.
- **Red:** Incorporates patient fears, beliefs, and trust factors.
- **Black:** Flags system overloads from population pressure.
- **Yellow:** Quantifies gains from reproductive health and family planning.
- **Green:** Innovates mobile clinics and care models in underserved areas.
- **Blue:** Coordinates multiagent systems across clinics, telehealth, and community workers.

**Monetization Gains:** Reduced maternal/child care costs; economic benefits from healthier populations; ROI in family planning interventions.

## SDG 4: Quality Education

**Business Problem:** Youth bulges overrun education systems; quality deteriorates under population pressure. **Six Thinking Dimensions:**

- **White:** Tracks student-to-teacher ratios and school capacity.
- **Red:** Gauges student, teacher, and community sentiment.
- **Black:** Highlights risks of dropout, overcrowding, and diminished outcomes.

- **Yellow:** Demonstrates value of optimal class sizes and family-size education policies.
- **Green:** Creates scalable e-learning and peer-to-peer learning agents.
- **Blue:** Coordinates across ministries, schools, and ed-tech platforms.

**Monetization Gains:** Higher productivity from educated workforce, reduced social support costs, and scalable ed-tech models.

## SDG 5: Gender Equality

**Business Problem:** Overpopulation exacerbates gender disparities—especially in education, reproductive autonomy, and economic participation. **Six Thinking Dimensions:**

- **White:** Integrates demographic, education, and empowerment metrics.
- **Red:** Captures societal and personal values around gender norms.
- **Black:** Identifies risks of gender-based violence and exclusion.
- **Yellow:** Shows value of female empowerment via family planning, education, and labor participation.
- **Green:** Designs mentorship and digital safe-space platforms.
- **Blue:** Aligns agents across health, education, and policy initiatives.

**Monetization Gains:** Economic growth from increased female labor force; reduced social costs from inequality.

## SDG 6: Clean Water and Sanitation

**Business Problem:** A growing population strains water resources and sanitation systems. **Six Thinking Dimensions:**

- **White:** Gathers water usage, population, and sanitation coverage data.
- **Red:** Understands user needs and behavioral patterns.
- **Black:** Models risk of contamination and resource depletion.
- **Yellow:** Calculates saved costs through efficient planning and family-size choices.
- **Green:** Develops water-saving tech and scalable sanitation agents.
- **Blue:** Coordinates across utilities, communities, and NGOs.

**Monetization Gains:** Infrastructure savings, reduced public health costs, and pay-per-use sanitation systems.

## SDG 7: Affordable and Clean Energy

**Business Problem:** Population growth increases energy demand, threatening clean energy access. **Six Thinking Dimensions:**

- **White:** Monitors population density and energy usage patterns.
- **Red:** Captures consumer sentiment about energy costs and preferences.
- **Black:** Anticipates power shortages or grid instability.
- **Yellow:** Quantifies ROI of renewable energy and demand-management interventions.
- **Green:** Invents microgrid and off-grid solutions.
- **Blue:** Coordinates agents across utilities, regulators, and consumers.

**Monetization Gains:** Energy efficiency savings, new microenergy markets, and reduced carbon penalties.

## SDG 8: Decent Work and Economic Growth

**Business Problem:** Large, young populations without sufficient job creation can cause unrest and underemployment. **Six Thinking Dimensions:**

- **White:** Analyses workforce demographics vs. job availability.
- **Red:** Gauges youth aspirations and worker sentiments.
- **Black:** Predicts unemployment and socioeconomic risk areas.
- **Yellow:** Shows dividends of education, family planning, and targeted job programs.
- **Green:** Designs digital job-matching and entrepreneurial platforms.
- **Blue:** Coordinates agents across business, education, and policy spheres.

**Monetization Gains:** Higher labor productivity, decreased welfare dependency, and vibrant digital economies.

## SDG 9: Industry, Innovation and Infrastructure

**Business Problem:** Population-led urbanization overloads infrastructure and hampers innovation scalability. **Six Thinking Dimensions:**

- **White:** Tracks population growth vs. infrastructure capacity.
- **Red:** Integrates resident sentiment and pain-points.
- **Black:** Flags infrastructural collapse risks.
- **Yellow:** Quantifies the value of sustainable infrastructure and population management.

- **Green:** Proposes modular, resilient infrastructure solutions.
- **Blue:** Synchronizes agents across urban planners, engineers, and communities.

**Monetization Gains:** Cost-efficient urban systems, smart city revenue models, and infrastructure longevity.

## SDG 10: Reduced Inequalities

**Business Problem:** Population growth often exacerbates income, gender, and geographic inequalities. **Six Thinking Dimensions:**

- **White:** Profiles inequities across demographic groups.
- **Red:** Captures community sentiment around fairness.
- **Black:** Models risk of social unrest or marginalization.
- **Yellow:** Shows returns from inclusive family planning and redistribution.
- **Green:** Generates agents to deliver services to underserved groups.
- **Blue:** Orchestrates across social protection, health, and education sectors.

**Monetization Gains:** Savings via targeted social programs and expansion of inclusive services.

## SDG 11: Sustainable Cities and Communities

**Business Problem:** Overcrowding, slums, and resource depletion threaten urban livability. **Six Thinking Dimensions:**

- **White:** Maps population density, land use, and infrastructure gaps.
- **Red:** Captures citizen perceptions of safety, housing, and settlement.
- **Black:** Predicts risks of disasters, overcrowding, and slum growth.
- **Yellow:** Quantifies gains from urban planning, population control, and clean infrastructure.
- **Green:** Designs modular housing, smart transport, and open-space interventions.
- **Blue:** Coordinates urban agents between city planners, communities, and utilities.

**Monetization Gains:** Smart city services, property value uplift, and service delivery efficiencies.

## SDG 12: Responsible Consumption and Production

**Business Problem:** Larger populations drive unsustainable resource consumption and waste. **Six Thinking Dimensions:**

- **White:** Tracks consumption patterns relative to population segments.
- **Red:** Gauges consumer values and behavior drivers.
- **Black:** Assesses risk of resource depletion and supply shocks.
- **Yellow:** Values low-impact consumption enabled by population stabilization.
- **Green:** Cocreates circular economy products and services.
- **Blue:** Coordinates agents across supply chains, manufacturers, and consumers.

**Monetization Gains:** Efficiency cost savings, circular economy revenues, and sustainable product markets.

## SDG 13: Climate Action

**Business Problem:** High population growth increases carbon emissions and climate pressure. **Six Thinking Dimensions:**

- **White:** Collates demographic and emissions data.
- **Red:** Captures community climate concerns and adaptation desires.
- **Black:** Models climate risks amplified by population growth.
- **Yellow:** Values low-carbon strategies, clean energy, and family planning.
- **Green:** Designs climate-positive systems—e.g., carbon sequestration linked to land use.
- **Blue:** Orchestrates agents across environmental policy, communities, and businesses.

**Monetization Gains:** Carbon credits, climate resilience funding, and sustainable impact investments.

## SDG 14: Life Below Water

**Business Problem:** Coastal population pressure drives overfishing, pollution, and habitat degradation. **Six Thinking Dimensions:**

- **White:** Monitors coastal populations and marine ecosystem health.
- **Red:** Captures fisherfolk sentiment and community dependence.
- **Black:** Flags risks of stock collapse and ecosystem imbalance.
- **Yellow:** Quantifies the value of sustainable fisheries and population management.

- **Green:** Innovates marine protection solutions and population-informed policy design.
- **Blue:** Coordinates agents across fisheries, policymakers, and communities.

**Monetization Gains:** Sustainable fisheries, marine ecotourism, and ecosystem service valuation.

## SDG 15: Life on Land

**Business Problem:** Population-driven land use leads to deforestation, habitat loss, and biodiversity decline. **Six Thinking Dimensions:**

- **White:** Tracks deforestation rates and population encroachment.
- **Red:** Captures rural community perspectives.
- **Black:** Models ecological collapse risks.
- **Yellow:** Values conservation tied to stabilizing populations.
- **Green:** Designs land use planning and rewilding systems.
- **Blue:** Coordinates across local communities, NGOs, and governments.

**Monetization Gains:** Ecosystem services, carbon trading, and conservation funding.

## SDG 16: Peace, Justice, and Strong Institutions

**Business Problem:** Youth bulges and population pressures can drive unrest, corruption, and institutional fragility. **Six Thinking Dimensions:**

- **White:** Monitors demographic pressures and governance metrics.
- **Red:** Captures societal trust and civic sentiment.
- **Black:** Assesses risks of conflict and institutional failure.
- **Yellow:** Demonstrates value of inclusive growth, education, and population planning.
- **Green:** Designs civic engagement platforms and youth empowerment agents.
- **Blue:** Coordinates agents across justice, civil society, and policy realms.

**Monetization Gains:** Reduced conflict costs, investment in stable institutions, and new governance technology services.

## SDG 17: Partnerships for the Goals

**Business Problem:** Achieving the SDGs requires collaboration across sectors, often hindered by misaligned demographic priorities. **Six Thinking Dimensions:**

- **White:** Maps partner capacities, demographics, and goal overlaps.
- **Red:** Understands stakeholder motivations and values.
- **Black:** Identifies partner misalignment or risk of goal failure.
- **Yellow:** Quantifies the value of coordinated population-SDG interventions.
- **Green:** Designs collaboration platforms and shared-agent networks.
- **Blue:** Orchestrates across governments, NGOs, corporates, and communities.

**Monetization Gains:** Enhanced funding efficiency, cross-sector coinvestment, and impact-driven partnerships.

# Glossary

## 0–7

**1IR—First Industrial Revolution** (circa 1760–1840) Transition from agrarian and manual labor economies to mechanized manufacturing. Driven by steam power, coal-fueled engines, and the mechanical loom, it established the factory system and rail transport. *See also: 2IR, Industrial Revolutions.*

**2IR—Second Industrial Revolution** (circa 1870–1914) Marked by electricity, assembly lines, and mass production, it enabled advances in steel, chemicals, global transport, and communications. *See also: 1IR, 3IR, Industrial Revolutions.*

**3IR—Third Industrial Revolution** (circa 1960–2000), also called the Digital Revolution. Introduced automation, computing, and electronics, shifting economies toward knowledge-based industries. *See also: 2IR, 4IR, Industrial Revolutions.*

**4IR—Fourth Industrial Revolution** (circa 2010–present) Integration of cyber-physical systems, AI, IoT, and cloud computing. Characterized by the convergence of digital, physical, and biological systems. *See also: 3IR, 5IR, AI, IoT, Industrial Revolutions.*

**5IR—Fifth Industrial Revolution** (emerging 2025 onward) Human-centric innovation combining creativity with AI-powered agents. Focus on collaboration, sustainability, and augmentation beyond automation. *See also: 4IR, 6IR, Agentic AI, AI-Powered Agentic Swarms.*

**6IR—Sixth Industrial Revolution** (visionary, 2035+) Expected rise of autonomous, self-improving agentic systems across physical, digital, and biological domains. Involves sovereign AI, quantum AI, and bio-digital integration. *See also: 5IR, 7IR, Agentic AI, Quantum AI.*

© The Author(s), under exclusive license to APress Media, LLC, part of Springer Nature 2026
A. F. Vermeulen, *Agentic Hyper-Personalized Dimensions*,
https://doi.org/10.1007/979-8-8688-1877-6

**7IR—Seventh Industrial Revolution** (circa 2045–2080) Envisions convergence of biological, artificial, and quantum intelligence. Focuses on the symbiosis of humans, machines, and ecosystems with ethical, ecological, and philosophical dimensions. *See also: 6IR, AGI, Agentic AI, Industrial Revolutions.*

# A

**AGI—Artificial General Intelligence** Highly autonomous systems with cognitive capabilities comparable to humans, able to transfer learning across domains. *See also: AI, Agentic AI, 7IR.*

**AI—Artificial Intelligence** Simulation of human intelligence in machines that learn, reason, and act. Includes narrow AI applications as well as progress toward AGI. *See also: AGI, Agentic AI, Generative AI.*

**AI-Powered Agentic Swarms** Collectives of autonomous, intelligent agents working collaboratively to achieve complex goals through distributed decision-making and adaptive learning. Each agent acts with autonomy, goal-setting, and collaboration. *See also: Agentic AI.*

**Agentic** The capability of an entity, particularly AI, to act with autonomy, intentionality, and self-direction. *See also: Agentic AI, AI-Powered Agentic Swarms.*

**Agentic AI** A class of AI systems designed with autonomy, self-direction, and adaptive decision-making. Unlike narrow AI, Agentic AI can set goals, reason contextually, and collaborate with humans and other agents. *See also: AI, AGI, Polymorphs, AI-Powered Agentic Swarms.*

# C

**CAGR—Compound Annual Growth Rate** A measure of the mean annual growth rate of an investment or metric over a specified multiyear period. *See also: Value.*

# D

**Dark Data** Data collected but not analyzed or used for decision-making. Often unstructured (e.g., logs, emails, sensor feeds), representing both hidden risks and untapped opportunities. *See also: Light Data.*

# G

**Generative AI**  A subset of AI focused on creating novel content, text, images, code, or simulations by learning from existing patterns. *See also: AI, Agentic AI.*

# L

**Light Data**  Structured, accessible information actively used in organizational processes. Typically well-governed and stored in enterprise systems. *See also: Dark Data.*

# P

**Polymorphs**  Autonomous AI agents within Fountain Intelligence that execute processes, continuously learn, and adapt. They serve as building blocks of larger workflows and collaborate in swarms. *See also: Agentic AI, AI-Powered Agentic Swarms.*

# R

**R-A-P-T-O-R**  Retrieve-Assess-Process-Transform-Organize-Report:  a  six-step processing framework for transforming raw or Dark Data into structured insights. *See also: Light Data.*

# V

**Validity**  Whether data accurately represents what it is intended to capture and is contextually appropriate for its use. *See also: Veracity, Value.*

**Value**  The degree to which data can generate actionable insights, benefits, or business outcomes. *See also: CAGR, Validity.*

**Variety**  The diversity of data formats, sources, and structures encompasses a wide range of data types, including text, images, video, logs, and sensors. *See also: Volume, Velocity.*

**Velocity**  The rate at which data flows into and out of systems often requires real-time or near-real-time processing. *See also: Volume, Variety, Viscosity.*

**Veracity**  The quality, reliability, and integrity of data, ensuring it is trustworthy and fit for decision-making. *See also: Validity.*

**Volume**  The overall quantity of data collected and stored across light and Dark Data sources. *See also: Variety, Velocity.*

**Vagueness**  Ambiguity or lack of clarity in data meaning and interpretation often leads to misinterpretation. *See also: Veracity, Vocabulary.*

**Volatility**  The lifespan of data reflects how long it remains valid, accurate, or useful before expiring. *See also: Velocity, Volume.*

**Visualization**  The process of representing data visually to enhance human understanding and decision-making. *See also: Value.*

**Virality**  The speed and extent to which data spreads across networks, particularly in social or digital contexts. *See also: Velocity.*

**Viscosity**  Friction in the movement or processing of data, often due to latency, system bottlenecks, or inefficiencies. *See also: Velocity, Volume.*

**Vocabulary**  The semantics, taxonomies, and metadata structures are required to interpret and integrate data consistently. *See also: Vagueness, Variety.*

**Venue**  The location and context of data sources, storage, and processing, such as cloud, edge, or on-premises systems. *See also: Volume, Velocity.*

# Index

GPSR Compliance
The European Union's (EU) General Product Safety Regulation (GPSR) is a set of rules that requires consumer products to be safe and our obligations to ensure this.

If you have any concerns about our products, you can contact us on

ProductSafety@springernature.com

In case Publisher is established outside the EU, the EU authorized representative is:

Springer Nature Customer Service Center GmbH
Europaplatz 3
69115 Heidelberg, Germany